KU-320-249

The Construction of New Buildings Behind Historic Facades

The Construction of New Buildings Behind Historic Facades

David Highfield BSc (Hons), MPhil, MCIOB

UNIVERSITY OF WOLVERHAMPTON
LIBRARY
Acc No. 878547
CLASS 690.24 HIG
CONTROL
DATE -3 NOV 1995
SITE RS

E & FN SPON
An Imprint of Chapman & Hall
London · New York · Tokyo · Melbourne · Madras

UK	Chapman and Hall, 2–6 Boundary Row, London SE1 8HN
USA	Van Nostrand Reinhold, 115 5th Avenue, New York NY10003
JAPAN	Chapman and Hall Japan, Thomson Publishing Japan, Hirakawacho Nemoto Building, 7F, 1–7–11 Hirakawa-cho, Chiyoda-ku, Tokyo 102
AUSTRALIA	Chapman and Hall Australia, Thomas Nelson Australia, 480 La Trobe Street, PO Box 4725, Melbourne 3000
INDIA	Chapman and Hall India, R. Seshadri, 32 Second Main Road, CIT East, Madras 600 035

First edition 1991

© 1991 David Highfield

Typeset in 11/13pt Baskerville by Photoprint
9–11 Alexandra Lane, Torquay, Devon
Printed in Great Britain at the University Press, Cambridge

ISBN 0 419 15180 X 0 442 31249 0 (USA)

Apart from any fair dealing for the purposes of research or private study, or criticism or review, as permitted under the UK Copyright Designs and Patents Act, 1988, this publication may not be reproduced, stored, or transmitted, in any form or by any means, without the prior permission in writing of the publishers, or in the case of reprographic reproduction only in accordance with the terms of the licences issued by the Copyright Licensing Agency in the UK, or in accordance with the terms of licences issued by the appropriate Reproduction Rights Organization outside the UK. Enquiries concerning reproduction outside the terms stated here should be sent to the publishers at the UK address printed on this page.

The publisher makes no representation, express or implied, with regard to the accuracy of the information contained in this book and cannot accept any legal responsibility or liability for any errors or omissions that may be made.

British Library Cataloguing in Publication Data

Highfield, David *1947–*
Construction of new buildings behind historic facades.
1. Buildings of historical importance. Conversion
I. Title
690.24

ISBN 0–419–15180–X

Library of Congress Cataloging-in-Publication Data

Available

To Chris, Sarah and Katherine

Contents

Preface

The construction of new buildings behind historic facades, better known as facade retention or 'facadism', is a unique phase in the history of architecture. The major surgery that it entails can only be applied to traditionally constructed buildings with loadbearing external envelopes and will not therefore be possible with the thin walled, framed buildings of today. Facade retention is also unique in that it has presented designers and constructors with special technical problems requiring new solutions, and highly specialized technical expertise that are unrelated to the well tried and tested methods used for the construction of new buildings. The building types most commonly associated with facade retention are usually dated between 1850 and 1940 and located in or near the centres of towns and cities. They are generally medium or large sized commercial and industrial buildings from two to six storeys in height and include banks, offices, shops, mills, warehouses and factories.

In addition to the technical challenges presented by the construction of new buildings behind historic facades, the subject also raises many philosophical and ethical questions. The drastic methods employed have caused much controversy amongst conservationists, architects, developers and planners. Some argue that historic buildings should only be retained in their entirety, while others accept facade retention as being a realistic and necessary compromise where there is a conflict of interests between conservationist and developer.

The use of facade retention, as a means of successfully combining architectural conservation with new development started to become common during the mid 1970s, although there were isolated examples before this, the best known being the elegant Nash Terraces in Regent's Park, London. During the 1980s, as the advantages of facade retention became more widely recognized by architects and developers, its use became more extensive until, at the present time, many examples can be seen in the process of construction in most of our towns and cities. There is still a vast stock of traditional buildings with valuable exteriors, but undistinguished interiors, that lend themselves to facade retention development, and it is anticipated that this drastic but highly successful and widely accepted means of re-using historic buildings will continue to increase in the foreseeable future. Further afield, facade retention has also become widespread in many countries with buildings that lend themselves to this form of re-use. In the USA, prime examples include Amussens Jewellery Store and the ZCMI building in Salt Lake City, and the Army and Navy Club and Bond Building in Washington DC. In Europe, one of the most recent and notable examples involved the retention of a major facade of the internationally renowned Louvre palace in Paris.

The aim of this book is to give a detailed insight into all of the key issues associated with facade retention, including: the background to, and reasons for, its widespread use; its acceptability as a means of architectural conservation; planning procedures, design, and the structural problems associated with this drastic form of building re-use, and the technical solutions used to overcome them. It therefore comprises an essential and invaluable source of information to all of those involved with facade retention projects, including building owners, developers, architects, surveyors, planners, conservationists, structural engineers and building contractors.

The financial implications of facade retention, and comparative costings with alternative solutions, have not been included in the book due primarily to the fact that, in the vast majority of schemes, the facade is retained because there is no legal choice owing to reasons of conservation, and therefore cost comparisons would only be of academic interest. Cost comparisons between facade retention and demolition and newbuild are also of limited value, since the constraint of having to retain a facade almost inevitably results in a significantly lower floor area than can be achieved by total demolition and newbuild, therefore preventing a true comparison of like with like.

In general, it is well known that facade retention is considerably more expensive than the many less drastic forms of building re-use, and that the structural and logistical complexities of building behind a retained facade usually make it more expensive than total demolition and newbuild. However, the only means by which a client or developer can achieve an accurate indication of comparative costings is by commissioning a detailed cost appraisal which must include the long- as well as short-term financial implications of opting for a range of different solutions.

The first chapter gives an overview of the re-use and conservation of historic buildings and explains the role of facade retention within the wider context of building redevelopment.

Chapter 2 explains the reasons why developers opt for facade retention, including the advantages to be gained, together with a range of other more general reasons for its widespread use.

Chapter 3 deals with the philosophical implications of facade retention and explains why, despite the 'purist' viewpoint, it is almost universally accepted by developers, architects, planners and conservationists as a means of re-using historic buildings. I would like, at this point, to acknowledge the contributions made to the preparation of this chapter by a number of major national conservation bodies, including The Royal Fine Arts Commission, The Cockburn Association, Edinburgh, The Scottish Civic Trust, The Victorian Society, The Georgian Group, The Society for the Protection of Ancient Buildings and SAVE Britain's Heritage.

Chapter 4 examines the formal planning authority procedures used when facade retention schemes are submitted for approval, and includes detailed consideration of the relevant legislative framework. I would like, at this point, to express my thanks to Trevor Houseago, Principal Planning Officer (Conservation), Leeds City Council, for his valuable help and advice in the preparation of this chapter.

Chapter 5 gives a set of specific design criteria and guidelines, developed from a wide ranging study of completed schemes, which are generally desirable as a basis for achieving architecturally acceptable designs.

Chapter 6 examines the principal technical problems met in the design and construction of facade retention schemes, and explains the range of solutions that can be used to overcome them. Aspects considered include providing temporary support to the retained facade, tying the facade back to the new structure, differential settlement and foundation design.

The final chapter contains illustrated descriptions of eight typical facade retention case studies, giving a detailed insight into each scheme and the techniques used to solve the various technical problems. The text is supplemented by a large number of photographs and detailed drawings and I wish to thank Sean Acaster for his painstaking work in preparing these drawings for final publication.

David Highfield

1

The redevelopment and conservation of historic buildings

1.1 Redevelopment of historic buildings

The need to provide new buildings in our towns and cities is often aggravated by the lack of open sites on which to carry out development. Established centres can seldom offer undeveloped sites and, in order to provide new accommodation, developers must inevitably focus their attention on existing buildings. The changing needs of society, together with major changes in commerce and manufacture, have meant that large numbers of older buildings have become redundant or obsolete while still possessing obvious, identifiable qualities. Such buildings, particularly if they are architecturally attractive and structurally sound, are often ideal for redevelopment. The redevelopment can be executed in varying degrees, ranging from the least drastic option of 'low-key' rehabilitation where all or most of the existing building is retained, with finishes and services merely upgraded, to the most drastic option of total demolition followed by the construction of an entirely new building.

1.2 The scale of redevelopment options

In practice, numerous redevelopment options exist and their relationships can be shown on a scale which ranges from the least to the most drastic.

1. Retention of the entire existing building structure, together with its internal sub-divisions, and upgrading of interior finishes, services and sanitary accommodation. In the most low-key of rehabilitation schemes existing stairs would be upgraded in preference to installing lifts, and simple heating systems would be used in conjunction with natural ventilation.
2. Retention of the entire existing external envelope, including the roof and most of the interior, with minor internal structural alterations and upgrading of interior finishes, services and sanitary accommodation. The structural alterations might involve the demolition of some interior sub-divisions or the insertion of new staircases and possibly lift shafts.
3. Retention of the entire existing external envelope, including the roof, with major internal structural alterations and upgrading of finishes, services and sanitary accommodation. The major internal structural alterations might include the insertion of new reinforced concrete stairs, lift installations, extensive demolition of interior structural walls, or the insertion of new floors where the original storey heights permit.
4. Retention of all the building's envelope walls and complete demolition of its roof and interior, with the construction of an entirely new building behind the retained facade. This option might occur with an isolated building where the entire external facade walls are worthy of retention, but where the developer requires totally new accommodation, unconstrained by existing internal elements.
5. Retention of only two or three elevations of the existing building and complete demolition of the remainder, with the construction

of an entirely new building behind the retained facade walls. This option might occur where the building is situated on a corner or end-of-block site.

6. Retention of only one elevation, a single facade wall of the existing building, and complete demolition of the remainder with the construction of an entirely new building behind the retained facade. This option might occur where the building has only one important facade, which is the main street elevation adjoining buildings on each side.
7. The most drastic redevelopment option would be to totally demolish the existing building and replace it with a new building.

Options 4, 5 and 6 all involve the construction of a new building behind the existing elevation(s), and fall within the category of 'facade retention'.

1.3 The choice of redevelopment option

In practice, the redevelopment option chosen will depend upon a variety of economic, legislative and other constraints. For example, option 1 on the scale, which can be described as 'low-key' rehabilitation, may be desirable because it is usually much cheaper than any of the other options, and it can produce the 'new' accommodation in a much shorter time. On the other hand, option 1 may be the least desirable because of the design and constructional limitations imposed by having to retain all of the existing structure and having to make it comply with current regulations, especially with regard to fire.

1.4 The effects of listed building legislation

One of the most important legislative constraints affecting the choice of redevelopment option involves the statutory protection of buildings of architectural or historic importance. Section 1 of the Planning (Listed Buildings and Conservation Areas) Act 1990 requires the Secretary of State for the Environment to compile lists of buildings of special architectural or historic interest in order that such buildings can be protected from demolition or insensitive alterations and therefore be preserved for the enjoyment of present and future generations. When a building has been included in the list of buildings of special architectural or historic interest, it is an offence to carry out complete or partial demolition without receipt of listed building consent from the relevant local authority. In deciding whether or not to grant consent for total or partial demolition of a listed building, the local authority must consult national conservation bodies, and consider any representations made by local conservation groups, together with the recommendations of its own internal professional conservation officers. Listed building legislation is discussed more fully in Chapter 4.

1.5 Facade retention as a means of conservation

The important contribution made by listed buildings to the quality of our towns and cities, together with their architectural and historic importance, has resulted in resistance by conservationists to their demolition and alteration. Thus, where a listed building is the object of proposed demolition or alteration in order to provide new accommodation, a conflict of interests will almost inevitably occur between the developer and the conservationist. The number of listed buildings in Great Britain at present is around 500 000 and it is therefore likely that such a conflict of interests will occur if the proposed development is within a historic town or city. This often results in some form of compromise which permits alteration and modernization of the existing building, provided that those elements considered worthy of preservation are retained as part of the overall scheme. Many listed buildings, in addition to their valuable exteriors, possess major internal features of architectural or historic importance, and must therefore be retained in their entirety. However, since the majority of listed buildings owe their status to the value of their exteriors only, one of the most common compromises involves the complete demolition of the building's interior and the erection of a new structure behind its retained facade.

2
Why facade retention?

2.1 General

Facade retention is a drastic form of building rehabilitation to which some of the important advantages associated with 'low-key' rehabilitation do not apply.

1. Most 'low-key' rehabilitation is cheaper than total demolition and newbuild, whereas the structural complexities of facade retention often make it more expensive.
2. 'Low-key' rehabilitation takes considerably less time to complete than total demolition and newbuild, with corresponding economic advantages, whereas facade retention usually takes longer.

However, there are many advantages to be gained from, and reasons for, constructing new buildings behind historic facades which make it both economically and functionally viable as a means of conserving and re-using our older buildings.

The advantages of facade retention, followed by the other more general reasons for its widespread use, are given detailed consideration in the following sections.

2.2 The advantages of facade retention

2.2.1 The demand for prestigious buildings

Many organizations, such as insurance companies, banks, building societies and other financial institutions, often prefer to operate from attractive, prestigious historic buildings, and are prepared to pay higher sums to purchase or lease such buildings. The construction of new, modern accommodation behind a prestigious historic facade can therefore add considerably to the value of a building. The retained facade projects the 'image' required by the user, whilst the new accommodation it encloses provides a modern efficient working environment.

2.2.2 The availability of financial aid

Although, as stated in section 2.1, facade retention is often more expensive than both 'low-key' rehabilitation and newbuild, it may be possible to offset some of the costs by obtaining financial aid. Financial aid is not available for all rehabilitation and conservation schemes, but in many cases, for example where buildings of architectural or historic merit are concerned, or where jobs are being created, it may be possible to obtain substantial grants, or loans at reduced rates of interest, towards the cost of the work.

Detailed information on the numerous sources of financial aid can be found in certain publications listed in Further Reading at the end of the chapter.

2.2.3 Planning permission may not be required

Under the Town and Country Planning Act 1990, planning permission is required for **development**. However, 'the carrying out of works for the maintenance, improvement or other alteration of any building, being works which affect only the **interior** of the building or which **do not materially affect the external appearance** of the building . . .' does not constitute development and therefore does not require planning permission. Thus, if the scheme does not affect the exterior appearance of the building, which is often the case with facade retention, there may be no need for the developer to obtain planning permission, resulting in shortening of the development period, and a corresponding saving in costs.

However, even if the exterior appearance of the building is not affected, planning permission will still be required if a 'material change of use' occurs: The Use Class Order 1987 (Statutory Instrument 1987 No. 764) designates 16 different use classes, and any proposed change from one of these use classes to another will require planning permission. Thus, many facade retention schemes, even though they involve no change to the building's exterior appearance, will still require planning permission. There are, though, many examples of schemes which do not require planning permission and these include facade retention. Providing the exterior appearance of the building is not materially altered, and its Use Class remains the same, (for example where an old office building is gutted and new office accommodation constructed behind the retained, unaltered facade) then planning permission will not be required.

Where there is doubt regarding the need for planning permission, an application can be made to the local planning authority for a determination as to whether or not planning permission will be required.

2.2.4 The possibility of increasing floor area by the insertion of additional floors

In the majority of facade retention schemes it is possible to insert additional floors when the new structure is erected. Many of the buildings that are the subjects of facade retention schemes have storey heights well in excess of modern requirements, particularly at their lower levels, making the insertion of additional floors possible. The construction of a new mansard roof can often facilitate the addition of one or two extra floors above the original roof level (118–120 Colmore Row, Birmingham, Case study one, p. 44).

Therefore, it is quite normal for an existing four or five storey building to be increased by two storeys, giving a considerable increase in floor area, and, in turn, freehold or leasehold value.

Where additional floors are inserted, their edges, and any associated suspended ceilings, may conflict with the existing window openings, and in such cases their effect on the retained facade should be minimized by good design. Similarly, where a mansard roof is used to facilitate the addition of extra floors, the design should ensure that it does not visually dominate the retained facade. These and other design aspects are given more detailed consideration in Chapter 5.

2.2.5 The effects of plot ratio control

Plot ratio control, which was introduced by the Ministry of Town and Country Planning in 1948, is a device used by planners to restrict the amount of floor space provided in new buildings in relation to their site areas. For example, a plot ratio of 3:1 will restrict the floor area of a new building to three times the area of its site. One of the principal reasons for the introduction of plot ratio control was to limit the heights of buildings in towns and cities so as not to impair the amenity and development possibilities of surrounding sites and buildings. It is used by different planning authorities on an *ad hoc* basis to suit their own requirements and is most likely to be applied in the central areas of large towns and cities, where plot ratios are often restricted to between 3:1 and 5:1.

The application of plot ratio control in restricting the size of new developments often makes it advantageous to re-use existing buildings, rather than to demolish and replace them. For example, many older buildings were built to a higher plot ratio than is currently permitted by planners. Some may have a plot ratio as high at 7:1 in areas where the plot ratios for new development may well be restricted to 3 or 4:1. Thus, it is clear that the retention and re-use of such a building could result in the provision of around twice as much 'new' floorspace as would be permitted if it were demolished and replaced with a new building. This important advantage applies not only to the less drastic redevelopment options (section 1.2, options 1–3) but also to facade retention schemes where the advantage can often be taken further by the insertion of additional floors as discussed in section 2.2.4.

2.3 Other reasons for facade retention

In addition to the advantages of facade retention, there are numerous other reasons for its widespread use which, whilst they cannot be specifically classed as advantages, are equally valid.

2.3.1 *The availability of suitable buildings*

Advances in industry and commerce, together with the demand for a more sophisticated interior environment for both work and leisure, have led to large numbers of buildings becoming outdated, redundant or obsolete and this, in turn, has provided an abundance of buildings that are suitable for redevelopment. Examples include large numbers of textile mills in the north of England, old factory and warehouse buildings in industrial centres, outdated institutional buildings such as schools and hospitals, older office buildings, and churches. Many of these buildings, particularly if they are structurally sound and architecturally attractive, are ideal candidates for facade retention schemes.

2.3.2 *The constraints of listed building legislation*

Many of the older buildings that are suitable for redevelopment will have been listed under Section 1 of the Planning (Listed Buildings and Conservation Areas) Act 1990, because of their architectural or historic importance. In such cases, the building, or at least those elements which led to its being listed, must be preserved in any event unless there are exceptional circumstances. The majority of listed buildings are listed because of the value of their exteriors, relatively few being listed because of internal features worthy of retention. The result of this is that there are many thousands of listed buildings with relatively insignificant interiors, but which do possess valuable facades which, by virtue of the legislation, must be preserved in all but the most exceptional circumstances.

2.3.3 *The architectural value of the facade*

Retention is favoured where the building's facade represents a prime example of its architectural style or period. A well known facade retention scheme carried out for this reason was St Paul's House, Leeds (see Case study two, p. 59) which represents the best example of an Hispano-Moorish style of architecture in the UK. The whole external envelope was retained and extensively restored to survive as a unique example of its architectural style.

2.3.4 *The context of the facade*

Where the facade is an integral part of a larger, unified design and would be impossible to replace without destroying the unity and integrity of that design. An excellent example is Lloyd's Bank, Harrogate (Case study three, p. 71), the facade of which comprised two of the twelve units forming Cambridge Crescent, designed by George Dawson in 1880.

2.3.5 *The facade's contribution to the townscape*

Where retention of the facade would preserve the integrity and character of a particular street scene or part of the townscape. Many 'pockets' of townscape in historic towns and cities have not been designed to any unified plan, but have evolved over long periods to produce attractive vistas or areas which would be spoiled if any integral part were lost in favour of a modern development. If, therefore, the facade in question forms an integral part of such an area, it is likely that its retention would be desirable.

2.3.6 *Dilapidation of the existing interior due to neglect*

Where the building's facade is of value for one or more of the reasons in sections 2.3.2–5, but where its interior is so dilapidated as a result of neglect, vandalism or fire damage that it is beyond restoration or re-use.

2.3.7 *Destructive alteration of the existing interior*

Where the building's facade is of value but its original interior has been subjected to extensive destructive alteration throughout its life, which has left no internal features worthy of retention.

2.3.8 Structural weakness of the existing interior

Where the building's facade is of value and its interior is structurally sound but incapable of supporting the loads which would be imposed as a result of its proposed new use. For example, many older buildings have timber floors which are incapable of carrying the loads imposed by modern office equipment. Thus, where a building with timber floors is to be converted into offices, the existing floors will either have to be strengthened or replaced with concrete or steel. If the latter option is chosen, then complete replacement of the existing interior with a new structure (i.e. the facade retention option) could well be the most appropriate solution.

2.3.9 Unsatisfactory internal layout

Where the building's facade is of value, but its existing internal layout is unsatisfactory and would require extensive structural alterations in order to adapt it to suit the required new usage.

2.3.10 Compliance with fire regulations

Where the building's facade is of value, but where upgrading of the existing interior to comply with current fire regulations, including the provision of means of escape, would involve extensive alterations which would be severely detrimental to the existing fabric.

2.3.11 Economic viability

Where the building's facade is of value, but where retention of the existing interior and its upgrading to provide the new accommodation would involve excessive expenditure, and, on completion, would not result in an economically viable building.

2.3.12 The client's accommodation requirements

Where the client requires a prestigious exterior constructed from high quality materials incorporating the skilled craftsmanship and detailing usually present in valuable historic facades; but where modern, spacious, air-conditioned accommodation is also required and can only be provided by insertion of a new structure.

Further reading

Brand, C.M. and Williams, D.W. (1984) *Howell James' Notes on the Need for Planning Permission* (3rd edn), Oyez Longman, London.

Civic Trust for the North East (1980) *Guide to Grants and Loans for Conservation*, Civic Trust for North East.

Department of the Environment (1987) *Historic Buildings and Conservation Areas — Policy and Procedures*, Circular 8/87, HMSO, London.

Ellis, C., Allan, M., Hannah, M., Ratcliffe, J. and Rock, D. (1984) Funding for Construction, *Architects' Journal*, **180** (34 and 35), 32–92.

Highfield, D. (1982) *The Construction of New Buildings Behind Historic Facades: The Technical and Philosophical Implications*, M. Phil. thesis, University of York, York.

Highfield, D. (1983) Keeping up Facades, *Building*, **245**(39), 40–1.

Highfield, D. (1984) Building Behind Historic Facades, *Building Technology and Management*, **22**(1), 18–25.

Highfield, D. (1987) *Rehabilitation and Re-use of Old Buildings*, E. & F.N. Spon (An imprint of Chapman and Hall), London.

Great Britain (1990) *Town and Country Planning Act 1990*, ch. 8, HMSO, London.

Great Britain (1990) *Planning (Listed Buildings and Conservation Areas) Act 1990*, ch. 9, HMSO, London.

Great Britain (1972) *Town and Country Planning (Scotland) Act 1972*, ch. 52, HMSO, London.

Great Britain (1987) *Town and Country Planning (Listed Buildings and Buildings in Conservation Areas) (Scotland) Regulations 1987*, Statutory Instruments, 1987 No. 1529 (S.112), HMSO, London.

Great Britain (1990) *Planning (Listed Buildings and Conservation Areas) Regulations 1990*, Statutory Instruments, 1990 No. 1519, HMSO, London.

Great Britain (1987) *Town and Country Planning (Use Classes) Order 1987*, Statutory Instruments, 1987 No. 765, HMSO, London.

3

The acceptability of facade retention as a means of architectural conservation

3.1 General

The major surgery to which valuable historic buildings are subjected during facade retention schemes, and the resulting loss of their integrity, has caused much controversy among conservationists, architects, developers and planners. Some conservationists believe that if a building is worth retaining, it should be retained in its entirety, and that using part of its shell to conceal new accommodation is an extremely false solution. This view is counterbalanced by those who, it could be argued, take a more realistic view and accept that compromise is necessary and some destruction and loss inevitable if the needs of both the developer and the conservationist are to be satisfied. This chapter examines the views expressed by those who find themselves involved with facade retention schemes, together with the principal factors in favour of facade retention, in order to demonstrate its widespread acceptability as a means of architectural conservation.

3.2 The purist's viewpoint

The principal argument against facade retention, or any kind of drastic alteration to a historic building, expressed by some conservationists, is that it results in the loss of the building's architectural and historic integrity. This 'purist' view is based upon the argument that if a building is worth retaining it should be retained in its entirety, and that to conceal a new building behind an unrelated historic facade is a spurious and unacceptable solution. Purists believe that what lies behind a building's facade should be related to it both architecturally and historically; for example, a Georgian domestic facade should be used only to enclose the rooms that it was originally designed to enclose, and not, as occurs in some facade retention schemes, modern open-plan office accommodation.

3.3 The realist's viewpoint

The more realistic view, that some loss of a building's integrity may be inevitable if it is to be preserved, is widespread amongst conservationists, very few of whom take a truly purist attitude. It is recognized that, unless we wish to preserve the majority of historic buildings purely as sterile museum pieces, some loss of their integrity will be inevitable since, in most cases, the key to preservation lies in their adaptation to living, working buildings which can fulfil some current, viable purpose. Whilst believing the total retention of any historic building to be the most desirable solution, to be followed wherever possible, most conservationists concede that facade retention, representing as it does the most drastic violation of a building's integrity, may be a more practicable and realistic solution in many cases when other factors, such as the nature and condition of the existing interior, are taken into consideration.

3.4 The practical acceptability of facade retention

Facade retention has become acceptable for a number of important practical reasons, and the majority of those concerned with architectural conservation consider it to be justifiable in the following circumstances.

1. Where the existing interior is undistinguished or uninteresting, having no architectural or historic value.
2. Where the existing interior is in a poor or dilapidated condition and is therefore beyond repair or restoration.
3. Where the original interior has been subjected to extensive alteration and modification which place it beyond restoration.
4. Where the existing internal elements are structurally incapable of supporting the loads which would be imposed as a result of the proposed new use.
5. Where the existing internal arrangement is unsuitable for adaptation, and would require radical alteration in order to make any re-use viable.
6. Where upgrading of the building to conform with fire regulations would necessitate considerable alteration of the existing interior.
7. Where the cost of refurbishing the existing interior would be prohibitive, therefore ruling out the economic viability of the scheme.

Most of the above reasons, which have been discussed more fully in Chapter 2, are accepted by the majority of conservationists as justifying facade retention and are concerned with the practical aspects of re-using historic buildings. However, there are other philosophical arguments, less concerned with the practical implications, which oppose the purists' view and can be used to further justify facade retention as a means of conservation.

3.5 The townscape factor

As stated previously, the purist's viewpoint is based on the premise that, if a building is worth retaining, it should be retained as an integral whole, the facade being related both architecturally and historically to the interior that it encloses. However, many conservationists support the view that, because a building's facade and its interior fulfil entirely different functions, they need not be related in any way. The facade represents the 'public face' of a building, forming an important component of the townscape, whereas the interior represents the 'private face', being seen only by the building's occupants. Furthermore, the interiors and exteriors of buildings can be seen as forming 'inside rooms' and 'outside rooms'. The outside room in an urban environment usually has its walls formed by a complexity of buildings' facades and its ceiling by the sky. Inside rooms are formed by the interior wall, floor and ceiling surfaces of buildings. If a building is the subject of a facade retention scheme, its original inside rooms will be lost, but the character and nature of the outside room, of which its facade is a part, will remain unaffected.

This divorcing of the functions performed by the exteriors and the interiors of buildings, which can be used as a means of justifying facade retention, may be taken further when the relative importance of the inside and outside rooms are considered. The exteriors of buildings which form the walls of outside rooms are experienced by a vastly greater audience than their interiors, and are therefore of far greater importance in terms of architectural conservation. In this respect, therefore, the retention of a building's facade, usually the most valuable element of a historic building (section 2.3.2), will be the primary consideration.

The preservation of the outside room, or townscape, is widely recognized as one of the most important elements of architectural conservation and is supported by relevant legislation aimed at preserving or enhancing the character or appearance of areas of special architectural or historic interest. The legislation makes provision for the designation of such areas as conservation areas which, in an urban context, may range from whole town centres to squares, terraces or smaller groups of buildings. In the urban context, conservation area legislation is concerned not with the interiors of buildings, but with preserving their exteriors and the contribution they make to the townscape.

It is clear, therefore, that the townscape is a major factor in architectural conservation, and that facade retention can be a useful and viable means of ensuring that it is preserved.

3.6 The economic acceptability of facade retention

The preceding discussion has been concerned principally with the practical and philosophical acceptability of facade retention. An additional consideration, of equal, or possibly greater importance to developers concerns the economic implications. Any redevelopment, whether it involves providing a totally new building or the adaptation of an existing building, must be economically viable, and this will be a significant factor in determining the most suitable redevelopment option (section 1.2). To aid the choice of redevelopment option, a detailed cost appraisal of a number of options will be necessary. This appraisal will not only include the redevelopment costs, but also a long-term appraisal based upon expected rental income which will vary according to the quantity and quality of the accommodation provided.

If the actual redevelopment costs are considered in isolation, then generally, the less drastic forms of redevelopment, where a large part of the existing interior is retained, tend to be the cheapest and therefore the most viable in the short term. However, in the long term, the more drastic redevelopment options, which create entirely new accommodation behind the retained facade, are often the most viable since they more readily allow an increase in both the quantity and quality of the accommodation, which, in turn, result in a higher capital value and, where applicable, greater rental income.

Considering the quantity of accommodation, one of the principal arguments against internal refurbishment, and in favour of facade retention, advanced by developers is that the interiors of many historic buildings are uneconomically planned and wasteful in terms of usable space. Their entrance halls, staircases, corridors and toilet areas are often capacious, and it has been estimated that many prestige buildings, such as offices and banks, constructed prior to 1940, can have as much as 40% of their gross floor areas in wasted and unusable space when judged by current standards.

A further example of wasted space is the excessive ground floor storey heights common to many historic buildings, which, after structural alteration, can provide space for the insertion of an extra floor level. It is clear, therefore, that with many historic buildings, the quantity of usable floor space can be significantly increased by removing their interiors and inserting new, more efficiently planned accommodation behind their facades. Facade retention permits an increase in floor space by allowing more efficient planning at each floor level and, in most cases, the insertion of additional floor levels. The latter is very common in facade retention schemes and can be effected in four ways.

1. By excavating beneath the original structure and adding basement accommodation.
2. By inserting an additional floor within the original ground storey where the height of the latter permits.
3. By keeping the new structure's storey heights to a minimum to allow an additional floor (or floors) to be inserted between the ground and the original roof levels.
4. By providing additional floors above the original roof level behind a new mansard roof structure.

The economic viability of facade retention can not only be expressed in terms of the increased quantity of floor space it provides. In many cases, facade retention also allows the quality of accommodation to be improved, therefore increasing capital value and, where applicable, rental charges. Thus, in addition to the possibility of increasing the quantity of accommodation, the improvement in its quality is also used by developers to justify major alterations, including facade retention to historic buildings.

It is evident, therefore, that facade retention is economically acceptable, since it allows an increase in both the quantity and quality of accommodation and goes farthest in fulfilling the developers' principal motive of profitability. It should be appreciated by developers, however, that facade retention schemes are usually only economically viable in the medium to long term since the complex technical solutions required in their construction are much more costly than the methods used in less drastic forms of refurbishment. In many cases, facade retention may be the only means of producing a scheme yielding an acceptable level of profitability and, if it were not permitted, the building might remain redundant and fall further into decay, ultimately resulting in its total loss. This latter possibility is the developers' greatest lever which is often used to obtain consent for facade retention schemes.

Perhaps the final word concerning the often over-riding importance of economic considerations in conservation should be left to William Morris, founder of the first conservation body in this country, who, almost a century ago, observed that '. . . we of this Society . . . have often had to confess that if the destruction or brutification of an ancient monument of art and history was "a matter of money", it was hopeless striving against it.'

3.7 The acceptability of facade retention to conservationists

The major practical, philosophical and economic arguments concerning facade retention have now been considered in an attempt to give a clearer indication of its acceptability. However, the ultimate test must be the attitudes of those most closely involved with architectural conservation towards actual facade retention schemes. Research has shown that, in the majority of schemes, both local and national conservation organizations, most of whom exert a high degree of influence on final planning decisions, do not object to the principle of inserting new buildings behind historic facades. In many cases the conservation organizations have gone further and positively approved of facade retention where they have seen it as a means, often the only means, of preserving a valuable exterior.

Even greater influence on the final planning decisions for facade retention schemes is exerted by local authority professional conservation officers who are involved with the processing of applications for planning permission and listed building consent. As with conservation organizations, the majority of professional conservation officers, whilst generally preferring to see the retention of any historic building in its entirety, accept facade retention as an appropriate solution, particularly where retention of the whole would be impractical or uneconomic due to the nature and condition of the existing interior.

This acceptance of facade retention is reflected on a much wider scale through the increasingly large numbers of schemes that have been completed in recent years, the majority of which would not have received approval without the support of the conservation organizations and professional conservation officers concerned, all of whom exert considerable influence over the final planning decisions (Chapter 4).

The granting of Civic Trust awards and commendations to completed facade retention schemes, and the citing by the Scottish Civic Trust of such schemes as good examples of re-using historic buildings in its manual on New Uses for Older Buildings in Scotland, are further indications that those with an interest and influence in architectural conservation approve of facade retention.

The principal reason for the widespread acceptance of facade retention is that, although it results in destruction of the building's interior, it does not, if carried out with care, interfere in any way with the building's role in forming the townscape, a role which many argue is one of the most important in architecture. Conservationists, supported by relevant legislation, place considerable value upon the preservation of the townscape or 'outside room', and they accept facade retention because it fulfils this important requirement. Some conservationists reinforce this view by arguing that the exterior of a building, in performing its public function as part of the townscape, need not necessarily be historically or architecturally related to its interior, since the latter performs a very different and private function.

A further, important argument in favour of facade retention that has been advanced by developers and has received the sympathy of planners and conservationists, is that it is often the only means of achieving an economically viable re-use for many historic buildings. In such cases, if facade retention were not permitted, the buildings concerned could well be left unused, resulting in their entire loss through decay and dilapidation.

It is clear, therefore, that for a variety of reasons, facade retention has come to be widely accepted as an important means of preserving our architectural heritage for the enjoyment of future generations. Despite the 'purist' philosophy, which, whilst being well known, is only held rigidly by a very small minority, it is not generally considered wrong to demolish the undistinguished interior of a historic building and to use its attractive, architecturally and historically important facade to enclose well-designed and serviced modern accommodation. In this way, the integrity of the townscape is preserved and, at the same time, the facade performs another vital architectural function by once again offering protection from the elements to a living, working interior. It might be appropriate, in conclusion, to draw upon again the wisdom of William Morris, the founder of our conservation movement. It would be presumptuous to guess what his attitude towards facade retention might have been, but it could be argued that, in principle, the views he expressed in a reference to the alteration of buildings 'in early times' are as applicable to facade retention as they are to the kinds of alteration which he accepted were often a necessary part of any building's history:

> '. . . if ambition . . . pricked on to change, that change was of necessity wrought in the unmistakable fashion of the time; . . . but every change, whatever history it destroyed, left history in the gap, and was alive with the spirit of the deeds done midst its fashioning. The result of all this was often a building in which the many changes, though harsh and visible enough, were, by their very contrast, interesting and instructive and could by no possibility mislead.'

Further reading

Brandt, S. and Cantacuzino, S. (1980) *Saving Old Buildings*, Architectural Press, London.

Department of the Environment (1987) *Historic Buildings and Conservation Areas — Policy and Procedures*, Circular 8/87, HMSO, London.

Great Britain (1990) *Town and Country Planning Act 1990*, ch. 8, HMSO, London.

Great Britain (1990) *Planning (Listed Buildings and Conservation Areas) Act 1990*, ch. 9, HMSO, London.

Great Britain (1972) *Town and Country Planning (Scotland) Act 1972*, ch. 52, HMSO, London.

Great Britain (1987) *Town and Country Planning (Listed Buildings and Buildings in Conservation Areas) (Scotland) Regulations 1987*, Statutory Instruments, 1987 No. 1529 (S.112), HMSO, London.

Great Britain (1990) *Planning (Listed Buildings and Conservation Areas) Regulations 1990*, Statutory Instruments, 1990 No. 1519, HMSO London.

Highfield, D. (1982) *The Construction of New Buildings Behind Historic Facades: The Technical and Philosophical Implications*, M. Phil. thesis, University of York, York.

Lloyd, D., Fawcett, J. and Freeman, J. (1979) *Save the City* (2nd edn), Society for the Protection of Ancient Buildings, London.

Morris, W. (1877) *Manifesto of the Society for the Protection of Ancient Buildings*, SPAB, London.

Morris, W. (1900) *Architecture and History, and Westminster Abbey*, Longmans and Co., London.

Scottish Civic Trust (1981) *New Uses for Older Buildings in Scotland*, HMSO, Edinburgh.

4

The planning and decision-making process

4.1 General

Before a facade retention scheme can be allowed to proceed it must receive various statutory approvals from the relevant local authority, and, in view of the drastic and controversial nature of facade retention, it is essential that developers and their architects are fully conversant with the procedures involved in obtaining these statutory approvals.

The majority of historic buildings that are the subject of facade retention schemes are listed buildings or buildings in conservation areas, and the procedures used by local planning authorities to process applications proposing major alterations to such buildings are laid down by central government in various Acts of Parliament and Regulations (listed in Further reading at the end of the chapter). Facade retention schemes involving listed buildings must be granted listed building consent and, in the majority of cases, planning permission (section 2.2.3.) and the procedures laid down by central government have resulted in a standardized approach in processing applications which shows little variation between different planning authorities. These procedures are described below, commencing with the informal negotiations and terminating with the planning committee's final decision.

First, however, the background to listed building legislation is discussed since it is also essential that developers and their architects be familiar with this important constraint affecting the vast majority of facade retention schemes.

4.2 The background to listed building legislation

The need to preserve our architectural heritage by protecting buildings of special architectural or historic interest, either in their own right or as part of a group forming a pleasing external environment, has been reinforced during the past two decades by the introduction of legislation principally in the form of Acts of Parliament and Regulations. The main statutory provisions affecting architectural conservation are as follows:

- The Town and Country Planning Act 1990;
- The Planning (Listed Buildings and Conservation Areas) Act 1990;
- The Town and Country Planning (Scotland) Act 1972;
- The National Heritage Act 1983;
- The Planning (Listed Buildings and Conservation Areas) Regulations 1990;
- The Town and Country Planning (Listed Buildings and Buildings in Conservation Areas) (Scotland) Regulations 1987.

4.2.1 The listing of buildings

Section 1 of the Planning (Listed Buildings and Conservation Areas) Act 1990 requires the Secretary of State to compile lists of buildings of special architectural or historic interest for the guidance of local planning authorities. Before buildings are included in the lists, the

Secretary of State must consult appropriate persons who have a special knowledge of, or interest in, such buildings. Lists are compiled by the Secretary of State by reference to national criteria, with selection being based upon the following principles.

1. All buidings constructed before 1700 which survive in anything like their original condition are listed.
2. Most buildings constructed between 1700 and 1840 are listed, though selection is necessary.
3. Buildings constructed between 1840 and 1914 are only listed where they possess definite quality and character and selection is designed to include the principal works of the principal architects.
4. Selected buildings constructed between 1914 and 1939 are listed where they are of high quality and are the work of the principal architects of that period.
5. A few outstanding buildings constructed after 1939 have been listed, and in 1987 the Secretary of State for the Environment introduced a new '30 year rule' allowing outstanding buildings that are older than 30 years to qualify for listing. This resulted in the listing of 18 post-war buildings, including Coventry Cathedral and the Royal Festival Hall, both of which were listed Grade I.
6. In addition to the above, an 'emergency' ruling also exists which allows buildings of architectural excellence over ten years old to be listed if they are threatened in any way.

In selecting buildings for listing, particular attention is paid to the following.

1. Special value within certain types, either for architectural or planning reasons or as illustrating social or economic history (e.g. industrial buildings, railway stations, schools, hospitals, theatres, town halls, markets, exchanges, almshouses, prisons, textile mills).
2. Technological innovation or virtuosity (e.g. cast iron, pre-fabrication or the early use of concrete).
3. Association with well-known characters or events.
4. Group value, especially as examples of town planning (e.g. squares, terraces or model villages).

Listed buildings are classified into grades according to their relative importance.

Grade I: Buildings of exceptional architectural or historic interest. Only about 2% of listed buildings are in this category.
*Grade II**: Particularly important buildings of more than special architectural or historic interest. About 4% of listed buildings are in this category.
Grade II: Buildings of special architectural or historic interest which warrant every effort being made to preserve them.

Although the responsibility for listing buildings rests with the Secretary of State, in practice a large proportion of listings is made on the initiative of local authorities and amenity groups. There are no statutory provisions for local authorities to be consulted before any building is listed, but suggestions from local authorities and local amenity groups interested in architectural conservation, for the inclusion of particular buildings in the lists, are encouraged.

Lists, giving details of buildings that have been listed, must be kept available for inspection free of charge at the office of the relevant local authority and at the National Monuments Record, 23 Saville Row, London. Lists of buildings within a particular local authority's area are usually kept in the office of the planning department and are available for inspection during normal office hours. The lists give the location of each listed building, together with details of its history and the architectural features which have led to its listing.

In Scotland, listed buildings are divided into Categories A, B and C.

Category A: Buildings of national or more than local architectural or historic importance, or fine, little-altered examples of some particular period or style. Approximately 7% of Scottish listed buildings are in this category.
Category B: Buildings of primarily local importance or major but altered examples of some period or style. About 70% of Scottish listed buildings are in this category.
Category C: Good buildings considerably altered; other buildings which are fair examples of their period; or buildings of no great

individual merit, but which group well with others in Categories A or B. About 23% of listed buildings are Category C.

4.2.2 Building preservation notices

Section 3 of the Planning (Listed Buildings and Conservation Areas) Act 1990 gives local authorities the power to serve Building Preservation Notices in respect of buildings which are not listed, but which they consider to be of special architectural or historic interest. A Building Preservation Notice becomes effective immediately it has been served and remains in effect for six months. The purpose of the Building Preservation Notice is to give immediate protection to a building and to allow time for the Secretary of State to include it on the statutory list if, after investigation, it is considered worthy of listing. The most common application of the Building Preservation Notice is where an unlisted building, which might be worthy of listing, is threatened with demolition. The serving of such a Notice will give immediate protection to the building, whereas the normal listing procedures would be too lengthy to prevent the demolition taking place.

4.2.3 The effects of listing

Once a building has been listed, or is the subject of a Building Preservation Notice, it is an offence to carry out unauthorized works of demolition (deemed also to include partial demolition), alteration or extension in a way which would affect its character. Works are authorized only if listed building consent has been obtained and the works are carried out in accordance with the terms of the consent and any conditions attached to it and, in the case of demolition, notice of the proposal has been given to the Royal Commission on Historical Monuments. Following the granting of listed building consent to demolish, the Commission must be given at least one month to make a record of the building if it wishes.

4.2.4 Obtaining listed building consent

The procedure for obtaining listed building consent is set out in section 10 of the Planning (Listed Buildings and Conservation Areas) Act 1990, and applications must be made to the relevant planning authority which is required to:

1. advertise any applications which it receives for listed building consent;
2. display a notice on, or near, the site to which the application relates (unless the application is for consent to carry out works which affect only the interior of a Grade II (unstarred) building);
3. take account of any representations received (e.g. from individuals, local amenity groups etc.) when the application is being considered.

Where the application is for consent to demolish (or partially demolish) a listed building, the planning authority must also formally notify the following national amenity societies: The Ancient Monuments Society, the Council for British Archaeology, The Georgian Group, The Society for the Protection of Ancient Buildings, The Victorian Society, and the Royal Commission on the Historical Monuments of England. Any representations received in response to these notifications should be taken into account when the application is being considered. Where the application is for consent to alter, extend or demolish any Grade I or II* building outside Greater London, or any grade of listed building within Greater London, the planning authority must also notify English Heritage.

Outside Greater London, the local planning authority is required to notify the Secretary of State of any application for listed building consent which it proposes to grant. This is to enable the Secretary of State to decide whether or not to call in the application for consideration (section 4.3.10). In Greater London, the planning authority is required to notify English Heritage of any application for listed building consent which it does not propose to refuse, and the authority cannot grant consent until English Heritage either authorizes it to do so or directs it how to determine the application.

These latter requirements do not, however, apply to all applications, particularly those which involve only minor demolition or alteration works. When notifications are sent to the Secretary of State (or, in Greater London, English Heritage) the planning authority should explain its decision, and enclose copies of any representations received, particularly those from the national amenity societies.

4.2.5 Conservation areas

Section 69 of the Planning (Listed Buildings and Conservation Areas) Act 1990 makes provision for the designation of conservation areas which are defined as 'areas of special architectural or historic interest, the character or appearance of which it is desirable to preserve or enhance'. Conservation areas may be large or small, from whole town centres to squares, terraces and smaller groups of buildings. The Planning (Listed Buildings and Conservation Areas) Act 1990 brings the demolition of buildings in conservation areas (whether listed or not) under control by applying, with modifications, the listed building control provisions of the Act. Anyone wishing to demolish (or partially demolish) an unlisted building within a conservation area must first apply to the local planning authority for conservation area consent, following which similar procedures to those involving listed buildings are put into effect.

4.3 Local Authority planning procedures

As previously stated there is little variation between the planning and decision making procedures practised by different local planning authorities when they consider proposals for major alterations to listed buildings.

Applications for planning permission and listed building consent are normally processed in a number of standardized stages, and these are described below.

4.3.1 Negotiations with the planning department

All local planning departments encourage architects to pursue informal negotiations with them as early as possible in the design process and before submission of formal applications for planning permission and listed building consent. This enables the planners to clarify their requirements and reduces the possibility of abortive design and waste of both the architect's and planning department's time. Negotiations, which tend to be more formal, also continue after the planning and listed building applications have been submitted, especially where controversial items still exist. These negotiations usually concern aspects of the design about which the planners and/or conservation groups are dissatisfied and normally result in either a successful defence by the architect, or a change in the design to remedy the contentious item.

4.3.2 Processing of the application

When the formal applications for planning permission and listed building consent have been received, they are registered and numbered before being allocated to a planning officer who organizes and co-ordinates all the subsequent advertisements, notifications and consultations and, finally, prepares the report for consideration by the planning committee. The planning officer who processes the application is usually from the development control section of the planning department. Development control officers specialize in planning standards and land use policies, including such aspects as plot ratio control, housing/office/industrial development etc., and possess no particular expertise in conservation or historic buildings, these aspects usually being dealt with by other specialist planning officers (section 4.3.6.).

4.3.3 Advertisement of the application

Details of the application are advertised in the local press and on the site itself. The advertisement must give a summary of the proposed development and details of availability of the application for public inspection and invites written representations to be submitted to the planning authority within a period of 21 days. The local planning authority is statutorily required to take account of any such representations received when the application is being considered.

Most planning departments usually also keep regularly updated lists which summarize all current applications and are available for public inspection during office hours. Many of the local amenity groups keep themselves informed of all relevant applications by regularly monitoring these lists.

4.3.4 *Notice of the proposals*

A notice giving details of the proposals must be posted on or near the site or building affected by the application.

4.3.5 *Notification to national amenity societies*

Where the application is for consent to demolish a listed building (which is deemed also to include partial demolition) as is the case with facade retention schemes, formal notice must be given to the following national amenity societies.

- ☐ The Ancient Monuments Society;
- ☐ The Council for British Archaeology;
- ☐ The Georgian Group;
- ☐ The Society for the Protection of Ancient Buildings;
- ☐ The Victorian Society;
- ☐ The Royal Commission on Historical Monuments of England;
- ☐ in certain cases, English Heritage must also be notified (see section 4.2.4).

The local planning authority is statutorily required to take account of any representations received from these bodies when the application is being considered.

4.3.6 *Specialist consultations within the local authority*

Specialist sections within the local authority are consulted and are required to report on their respective aspects concerning the application. The sections consulted include the highways and transportation department (or, in some authorities, the city engineer's department), which considers the effects of the proposals upon future road developments and other aspects concerning highways, the environmental health department and the conservation section. The Town and Country Planning (Listed Buildings and Buildings in Conservation Areas) Regulations, 1987 recommends planning authorities to employ officers with expertise in conservation to provide specialist advice on such matters. Most planning authorities have, on this recommendation, set up conservation teams, and it is they who are consulted by the planning officer processing the application to provide an important input into the total information which will be used to compile the final report to the planning committee.

4.3.7 *Consideration of development control implications*

The planning officer processing the application, if a development control officer, examines the aspects of the application affected by planning standards and land use policies. If, as occurs in some planning authorities, the processing officer is not from the development control section, these aspects will be the subject of an additional internal consultation similar to those described above.

4.3.8 *Preparation of the formal report*

The written responses to all of the external and internal consultations described above are used by the planning officer processing the application to prepare a written report giving the factual details and concluding with recommendations as to whether or not the application should be approved.

4.3.9 *Determination by the planning committee*

The written report is signed by the chief planning officer and presented to the planning committee, which uses it in determining whether or not to approve the application.

During the above processing of applications, informal and formal negotiations between the planners and the architect are continued where necessary. Aspects of the design with which the planning officers are dissatisfied, particularly with regard to conservation aspects, are discussed or formally notified to the architect with the aim of rectifying them, usually by design modifications, before preparation of the final report to the planning committee.

4.3.10 Referral of the application to the Department of the Environment

In addition to the above stages in the planning process, where a planning authority proposes to support an application for listed building consent involving a Grade I or II* building, or the demolition of a Grade II building, it must refer the application to the relevant regional office of the Department of the Environment. In doing so, copies of any representations received, particularly concerning matters of conservation, should also be enclosed. The purpose of this is to enable the regional office to advise the Secretary of State to 'call in' the application for the Secretary's own decision, where this is considered appropriate. The regional office, after considering the details of the application, must conclude either that the local authority be left to make its own decision or that the application be called-in. Generally an application will not be called-in where representations from national conservation bodies and local amenity groups do not object to the proposals. If an application is called-in, an enquiry, conducted by an inspector from the Department of the Environment, may be held. At the enquiry evidence is given by the applicant and the local authority, following which the inspector produces a report, which includes his/her own recommendations, and which is used by the Secretary of State or a senior officer to make the final decision.

4.4 The planning procedures in practice

It is clear from the general description of planning procedures in the previous section that specialist conservation officers within planning authorities, together with outside conservation groups, are provided with an opportunity to influence planning decisions concerning listed buildings. However, the actual extent to which they are allowed to influence final planning decisions depends on the practical application of the general planning procedures, and varies from authority to authority. The following paragraphs explain how the local authority planning procedures operate in practice and give an insight into the extent to which conservationists are actually allowed to influence the final planning decisions.

4.4.1 Processing of the application

Most local authority planning departments allocate the responsibility for processing applications to a development control officer. These officers specialize in the wider aspects of planning and have no specialist expertise in the field of conservation. The development control officer is responsible for using the representations made by the various external and internal consultees, including the specialist conservation officer, to prepare the final report to the planning committee and, having no specific expertise in conservation, might not give this aspect the attention that it clearly warrants where a listed building is affected.

In a few local authorities, however, applications affecting listed buildings are processed by a conservation officer in order that due consideration and emphasis on the conservation aspects is ensured in the final report to the planning committee. In the case of applications affecting listed buildings, this appears to be more appropriate since the conservation officer is equally capable of carrying out the other consultations and, at the same time, can see that the conservation aspect is given due attention in his final report.

4.4.2 The conservation officer's influence

The extent to which the specialist advice given by the conservation officer is reflected in the final report to the planning committee depends to some extent on the subjective judgement of the development control officer who prepares it. Although the officer is required to consider all of the representations received, their inclusion is dependent on the importance that he places on them. A further important factor that can affect the conservation officer's influence over the final planning decision is the attitude of the chief planning officer and his senior subordinates towards conservation. This can have an important effect in determining the extent to which the conservationists' views are allowed to influence the final planning decision.

4.4.3 Presentation of conservation advice to the planning committee

Very few local authority planning departments provide, as a matter of course, for a conservation officer to be present at the planning committee meeting to present the conservationists' case. Generally, the only professional officer present at the planning committee is usually either the chief planning officer or his deputy. Clearly, the specialist conservation officer's advice, together with the representations from outside conservation bodies, will carry far greater weight where the conservation officer is permitted to present evidence directly to the planning committee.

4.4.4 Conservation area advisory committees

In many planning authorities, an important additional means of formal co-ordinated representation for conservationists and other interested parties is provided through local conservation area advisory committees or similar bodies, which are set up on the advice of central government. This advice, first conveyed to local authorities in Ministry of Housing and Local Government Joint Circular 61/68, dated December 1968, is currently conveyed to them in paragraph 68 of Department of the Environment Circular 8/87, 'Historic Buildings and Conservation Areas — Policy and Procedure', March, 1987.

Conservation area advisory committees are set up, usually on the initiative of the planning authority, to give advice on applications affecting conservation areas, listed buildings and other related matters. Their membership is normally strongly biased in favour of those concerned with conserving the architectural heritage. Membership usually comprises representatives of all local amenity groups, together with planning officers (acting in an advisory capacity) and councillors. Conservation area advisory committees usually meet at least monthly, and their minutes must be made available to the planning committee where relevant. The degree to which conservation area advisory committees can influence planning decisions varies between authorities; for example, in some local authorities any advice given to the chief planning officer by the conservation area advisory committee on a specific issue must be included in the formal report to the planning committee on that issue. In other authorities, however, conservation area advisory committees may not be given this degree of influence.

4.5 Conclusion

It is clear from the preceding sections, which describe the planning procedures in detail, that all of those concerned with building conservation are given ample opportunity to state their case and exert influence on the final planning decisions. Central government requires that certain national conservation bodies be consulted where an application includes proposals for demolition (deemed also to include partial demolition) of a listed building and that their representations, together with those from local amenity groups, should be taken into account when considering the application. In addition, central government advises local planning authorities to set up teams of specialist conservation officers to advise on applications affecting historic buildings. The final decisions on applications affecting listed buildings are, therefore, subjected to the influence of external conservation bodies, both national and local, together with specialist officers inside the planning authorities. However, although the planning authorities process applications affecting listed buildings within the same general framework, there are notable variations in the ways in which some of them operate in practice, particularly regarding the extent to which their specialist conservation officers, and the external conservation groups, are actually allowed to influence the final planning decisions. In most planning authorities, the professional conservation officers who are employed for their expertise in the field of conservation, appear to take a secondary role in advising on applications affecting listed buildings. The extent to which their advice is allowed to affect the planning committee's decision is to some extent dependent upon the judgement of the planning officer who prepares the final report to the committee, and upon the attitude towards conservation of the chief planning officer. Given the importance that central government places upon building conservation, together with the significant strength of public opinion in its favour, it could be argued that the professional conservation officer should be given the key role, rather than an officer with no particular expertise in conservation. This would be more effective in ensuring that due stress was placed on the conservation aspects in the final report to the planning committee. However, the vast majority of local authority planning departments have an enlightened attitude towards architectural conservation and, despite their secondary role in the planning process, conservation officers do in practice exert a considerable degree of influence over final planning decisions where listed buildings are concerned.

With regard to the external conservation bodies and amenity groups, together with other interested parties, their degree of influence can be considerably enhanced where there exists some form of joint advisory panel on conservation. Planning authorities are advised by central government to establish conservation area advisory committees, mainly consisting of interested parties who are not members of the authority, and refer to them for advice applications which would affect the character or appearance of a conservation area. The majority of authorities have, on this recommendation, set up such committees, which also advise on applications affecting listed buildings. Usually their minutes must be notified to the planning committee and must be used, where relevant, in the preparation of reports to the committee. This, together with the fact that such committees bring together all those bodies with a concern for conservation to form a more powerful, unified group, significantly increases the influence of conservationists in determining planning decisions.

A further aspect of the planning process, which some developers have used as a basis for negotiation when major alterations to listed buildings are proposed, is the appeals procedure. A developer might, for example, make an application for a scheme that involves a greater element of demolition than is actually required, with the ultimate intention of gaining approval for a less drastic scheme following rejection of the initial application, on the premise that as a result of an appeal, the planning authority might be more open to negotiation and compromise having already rejected one proposal.

There are many examples of facade retention schemes receiving approval following initial rejection (and subsequent appeal) of applications to totally demolish listed buildings, and it is clear, therefore, that some developers have employed this tactic to their advantage, although it is likely that some of these schemes might have received approval without the need for the initial, more drastic, proposals.

Further reading

Bassett, B. (1979) *Conservation Area Advisory Committees*, Diploma in Conservation Studies dissertation, University of York, York.

Department of the Environment (1987) *Historic Buildings and Conservation Areas — Policy and Procedures*, Circular 8/87, HMSO, London.

Highfield, D. (1982) *The Construction of New Buildings Behind Historic Facades: The Technical and Philosophical Implications*, M. Phil. thesis, University of York, York.

Great Britain (1990) *Town and Country Planning Act 1990*, ch. 8, HMSO, London.

Great Britain (1990) *Planning (Listed Buildings and Conservation Areas) Act 1990*, ch. 9) HMSO, London.

Great Britain (1972) *Town and Country Planning (Scotland) Act 1972*, ch. 52, HMSO, London.

Great Britain (1987) *Town and Country Planning (Listed Buildings and Buildings in Conservation Areas) (Scotland) Regulations 1987*, Statutory Instruments, 1987 No. 1529 (S. 112), HMSO, London.

Great Britain (1990) *Planning (Listed Buildings and Conservation Areas) Regulations 1990*, Statutory Instruments, 1990 No. 1519, HMSO, London.

5

Design criteria for facade-retention schemes

5.1 General

Although the design of every facade retention scheme should be assessed on its own particular merits, the process of evaluation can be significantly aided by reference to certain criteria which are generally desirable as a basis for achieving acceptable schemes. The criteria that follow, which are intended to cover the main aspects and are therefore by no means exhaustive, have been developed from a combination of sources which include a set of guidelines on alterations to listed buildings compiled by the Historic Buildings and Monuments Commission for England (set out in Department of the Environment Circular 8/87 *Historic Buildings and Conservation Areas-Policy and Procedures*), the practical lessons learned from a number of completed facade retention schemes, and the findings of research into the subject. It should be made clear that the criteria are not concerned with the wider philosophical, practical and economic issues, discussed in Chapter 3, used to determine whether or not schemes are acceptable in principle. They are concerned only with the subsequent and more specific problems of actual design, and should therefore be of value to architects, planners and conservationists in the evaluation of facade retention schemes.

5.2 Alterations to the retained facade

Any alterations which might adversely affect the special architectural or historic interest of the retained facade, or its contribution to the townscape, should be kept to a minimum.

If a facade is considered worthy of preservation, it will generally be because of the contribution it makes as it stands, and, in view of the considerable licence given to developers by facade retention, any further alterations which may adversely affect it should be strongly discouraged. An example of where extensive alterations to a previously unspoiled facade have significantly reduced its former architectural merit and character is the Church Institute, Leeds, where the ground floor elevations have been almost totally replaced by obtrusive modern shopfronts. Figure 5.1 shows the original building, and Figure 5.2 shows the scheme after completion.

5.3 Concealment of the new structure

The completed scheme should conceal, as fully as possible and from all external viewpoints, the existence of the new structure behind the retained facade.

The ease or difficulty in achieving this will depend on the nature of the scheme. For example, concealment of the new structure will be relatively easy to achieve if the scheme involves the retention of only a single street elevation which forms part of a continuous group. On the other hand, if side elevations are involved, concealment will be more difficult.

It will be appreciated that, in some schemes, total concealment of

Figure 5.1 Church Institute, Leeds: before redevelopment.

Figure 5.2 Church Institute, Leeds: after redevelopment.

Figure 5.3 142–144, Princes Street, Edinburgh: before redevelopment.

Figure 5.4 142–144, Princes Street, Edinburgh: after redevelopment.

the new structure may not be possible, and in such cases its concealment from the principal viewpoints, such as main thoroughfares, should be given priority. Figures 7.9, 7.10 (page 5 and 56) 5.3 and 5.4 show how concealment of the new structure has been treated in two different schemes. At 118–120 Colmore Row, Birmingham (Case study one, p. 44), the existence of the new structure is successfully concealed from all viewpoints along the adjacent thoroughfare (although not from the upper storeys of surrounding buildings), whereas the design of the scheme at 142–144 Princes Street, Edinburgh, leaves no doubt that a new structure exists behind the retained facade.

5.4 Scale of interior spaces

The scale of the interior spaces immediately behind the facade should be in keeping with the use of the original building and, therefore, the design of the facade. In certain cases it will also be desirable for the interior lighting arrangements and furnishings to be in keeping with the use of the original building.

This is a vital requirement in many facade retention schemes since the character of, and the atmosphere created by, a retained facade can be greatly influenced by what is seen through its windows. A common example of poor design practice in this respect is the creation of large open plan office spaces with suspended ceilings and fluorescent lighting behind retained Georgian domestic facades. This treatment, particularly at night or on dull days when the lights are on, is one of the most blatant and insensitive ways of detracting from the character of a retained facade. One way of avoiding it, even if brightly lit open plan spaces are essential, is to locate a number of smaller offices, which may well be required for senior staff, immediately behind the facade and to use light fittings, finishes and furnishings that reflect its domestic nature.

5.5 New floor levels

The new floors should be inserted at similar levels to the existing so that their edges and, where applicable, their associated suspended ceilings, do not conflict with the existing window openings.

In some schemes this criterion may be difficult to fulfil, a typical example being where the ground storey height of the existing building is excessive, lending itself to the insertion of an extra floor, the edge of which must pass across the tall ground floor windows. In cases such as this, where new floors passing across the window openings cannot be avoided, it is essential that their effect on the retained facade is minimized by the use of other design features such as special glazing, opaque panelling etc.

5.6 Restoration of major original features

Where the retained facade has suffered adverse, destructive alterations resulting in the loss of important original features, the scheme should, if possible, incorporate an accurate restoration to its former design.

One of the most common forms of adverse destructive alteration to the exteriors of historic buildings is the insertion of shopfronts at ground level, and some of the most successful facade retention schemes have involved the restoration of facades which have been spoiled in this way. Two examples are 1–3 Baxters Place, Edinburgh (Case study five), Figures 7.63 and 7.64 (page 111), where undistinguished projecting Victorian shopfronts were removed, and the ground storey carefully restored; and Abbey House, Glasgow (Case study seven), Figures 7.87 and 7.88 (page 137), where large plate glass shop windows were removed and replaced by arcading re-created to match the original design.

In some cases, however, the building's proposed new use may require a modern ground level frontage, making restoration or retention of the original inappropriate. In this event, compliance with the following criterion will be desirable.

5.7 Design of new frontages

Where the incorporation of a modern ground floor frontage is

Figure 5.5 Barclays Bank, Manchester: after redevelopment.

Figure 5.6 Barclays Bank, Manchester: after redevelopment.

Figure 5.7 *St Andrew House, Leeds: after redevelopment.*

essential to the proposed new use, its design should be in sympathy with the rest of the elevation. In particular, it should respect the rhythm of the original as reflected by its fenestration and, where possible, should incorporate any existing ground floor details of interest.

Good examples of sympathetic modern frontages are illustrated by two schemes, both of which had large plate glass shopfronts prior to their re-development. At Lloyd's Bank, Harrogate (Case study three), Figures 7.34 and 7.35 (page 82) the new six-bay frontage is in keeping with the rhythm of the facade above with its six bays of windows and dormers. Similarly, at Barclay's Bank, Colmore Row, Birmingham (Case study eight), Figure 7.96 (page 147), the rhythm of the facade's fenestration is reflected by the new frontage which also has six original cast iron columns, with their ornate capitals, incorporated into its design.

5.8 New materials

Any new materials used in providing a modern frontage, or in the restoration or replacement of parts of the retained facade, should take account of the existing materials and match them in quality, texture and colour.

Ideally, the same materials as the original should be used to enable the new work to blend with the existing, as at 1–3 Baxter's Place, Edinburgh (Case study five), Figures 7.64 and 7.65 (pages 111 and 112) where both the ground floor frontage and the roof were restored. Failure to do this can result in an undesirable contrast between new and existing as occurred at Lloyd's Bank, Harrogate (Case study three), Figure 7.35 (page 82), where the dark, smooth, polished granite of the new frontage conflicts with the light, textured gritstone ashlar of the facade above.

5.9 Artificial materials

In certain circumstances, the use of artificial materials in restoration and replacement is both effective and acceptable.

Typical examples are where use of the original materials and craftsmanship are economically prohibitive, as at Abbey House, Glasgow (Case study seven), Figures 7.87 and 7.88 (page 137), where the restored arcading and capitals were re-created in glass reinforced concrete; and at St Paul's House, Leeds (Case study two), Figures 7.19 to 7.21 (pages 67 and 68), where pigmented glass reinforced plastic was used to re-create the original parapet and minarets which were formerly in terra-cotta.

5.10 Scale, appearance and position of extensions

If the scheme involves extending beyond the confines of the retained facade, the extension should not dominate the original in scale, appearance or position.

Many facade retention schemes involve extensions or additions, and the problem of harmonizing new with existing will vary with each individual scheme. For example, a rear extension, which cannot be seen with the retained facade, will present fewer problems than a side extension which must be viewed with the existing. In the latter case, the new work may be designed either to match and blend with the existing, or to contrast with it and thus be viewed as a separate building. The schemes at Barclay's Bank, Manchester, Figures 5.5 and 5.6; 142–144 Princes Street, Edinburgh, Figures 5.3 and 5.4; 3–13 George Street, Edinburgh (Case study six), Figure 7.77 (page 125); and St Andrew House, Leeds, Figure 5.7, illustrate the contrasting and sometimes insensitive approaches to the problem.

5.11 Retention and re-creation of secondary original features

Any features which are an essential part of the original building's design, character and function, for example windows, doorways, chimneys, area railings etc., which indicate the domestic nature of a terrace, should be retained or re-created in the new scheme, even when they are no longer required.

In many schemes, such features may be superfluous to the new use, but their retention will be essential if the true character of the retained facade is to be preserved. In such cases, windows and doors that are no longer required can be sealed off from the inside, leaving a dummy to the exterior, as at Regency House, Edgbaston, Birmingham, Figure 5.8, where only the third door from the right was required, the others being dummies. It will be noted, however, that in this example, whilst the doors and railings have been retained, a vital element of its original character has been lost by the omission of the chimneys. Although these would, like the doors, have been superfluous, the construction of dummy chimneys in the new roof would have successfully completed the preservation of the group's domestic character, as at 1–3 Baxter's Place, Edinburgh (Case study five), Figure 7.64 (page 111), where three dummy chimneys were built for this purpose.

Figure 5.8 Regency House, Edgbaston, Birmingham: after redevelopment.

5.12 Preservation or re-creation of the original roof

Where the existing roof is a dominant feature, its original shape, pitch, covering and ornament should be preserved or recreated where possible.

Figure 5.9 113–115, George Street, Edinburgh: after redevelopment.

The roof is often a dominant feature of a building; a visual extension of its facade, and its retention can therefore be as important as that of the facade itself. Figures 5.9 and 7.35 (page 82) show examples of the retention and re-creation of the original roofs on two schemes. At 113–115 George Street, Edinburgh, Figure 5.9, the existing roof was retained *in situ*; whilst at Lloyd's Bank, Harrogate (Case study three), Figure 7.35, a replica of the original roof was constructed to match those of the other buildings in the group.

Retention or re-creation of the original roof is not, however, always possible, since many facade retention schemes involve the addition of extra floors at high level, requiring the addition of a new roof which may extend above the facade. In such cases, the following criterion should be applied.

5.13 Design of new roofs

Where the provision of a new, modern roof structure, extending above the retained facade, is necessary, it should be designed so that it is not visible from street level viewpoints in order to avoid it detracting from the appearance and character of the retained facade.

This is commonly achieved by setting back and/or raking back the new roof from the plane of the retained facade in the form of a mansard. Such a solution, however, may only conceal the roof from those viewpoints which are relatively close to the building and if important distant viewpoints do exist, with sight-lines which bring the new roof into view, it is essential that its design be as restrained as possible so as not to detract from the retained facade. The following examples illustrate how good design can minimize the effect of modern roof structures on retained facades. The double mansard at 118–120 Colmore Row, Birmingham (Case study one), Figures 7.9 and 7.10 (page 56), is a good example of a modern roof which is not visible from any of the surrounding street level viewpoints. Figure 7.88 (page 137) shows the use of a 'restrained' modern roof design at Abbey House, Glasgow.

5.14 Visual and decorative features

Towers, turrets, spires, cupolas etc., which are important visual and decorative features, making an important contribution to the building's character and the townscape, should be retained or re-created wherever possible. Excellent examples of the re-creation of important decorative features, which had been removed during the buildings' lifetimes, occur at St Paul's House, Leeds (Case study two), Figure 7.20 (page 68), where the ornate parapet and five corner minarets were replaced in their entirety; and the Church Institute, Leeds, Figure 5.2, where the dominant central spire was re-constructed.

Further reading

Department of the Environment (1987) *Historic Buildings and Conservation Areas — Policy and Procedures*, Circular 8/87, HMSO, London.

Highfield, D. (1982) *The Construction of New Buildings Behind Historic Facades: The Technical and Philosophical Implications*, M. Phil. thesis, University of York, York.

6

The technical aspects of facade retention

6.1 General

Facade retention is a unique phase in the history of architecture. It can be applied only to traditionally constructed buildings with load-bearing external envelopes, and will not therefore be possible with the thin-walled, framed buildings of today when and if they become listed buildings in the future. Facade retention is also unique in that it has presented building designers and constructors with technical problems which are unrelated to the design and construction of new buildings and which require highly specialized solutions and technical expertise. The major surgery applied to historic buildings which are the subject of facade retention schemes differs greatly from the well tried and tested techniques of new construction and it is often further aggravated by the unpredictable nature of the buildings concerned.

Facade retention is now a widely accepted method of redeveloping existing buildings and it is envisaged that the technique will become more common in the future as the demand for modern accommodation increases and the availability of open sites decreases.

The principal technical problems common to most facade retention schemes include:

1. providing temporary support to the retained facade during demolition of the existing building's interior and construction of the new structure;
2. the installation of permanent structural ties to hold the retained facade back to the new structure;
3. making allowance for differential settlement between the new structure and the retained facade in order to avoid subsequent structural damage;
4. providing a foundation system to the new structure which will not impair the stability of the retained facade.

It should be made clear at this stage that in the majority of facade retention schemes the facade is retained as a non-load-bearing element of the new building, supporting only its own self-weight and receiving lateral support from the new structure.

6.2 Temporary support systems

In all buildings comprising load-bearing external walls, there is an inter-dependency between those walls and the elements they carry. Whilst the external walls provide structural support to many internal elements (floors, roof structure and some cross-walls), these internal elements, in turn, provide lateral support to the walls. Thus, when this lateral support to the external walls is totally removed by demolishing the building's interior, it becomes necessary to provide some means of temporary support to the facade until the new structure is constructed and the retained facade tied back to it. A major consideration in the design of a temporary support system is that it must provide the facade with stability and resistance against wind-loads from both sides, to which it will be subjected for an

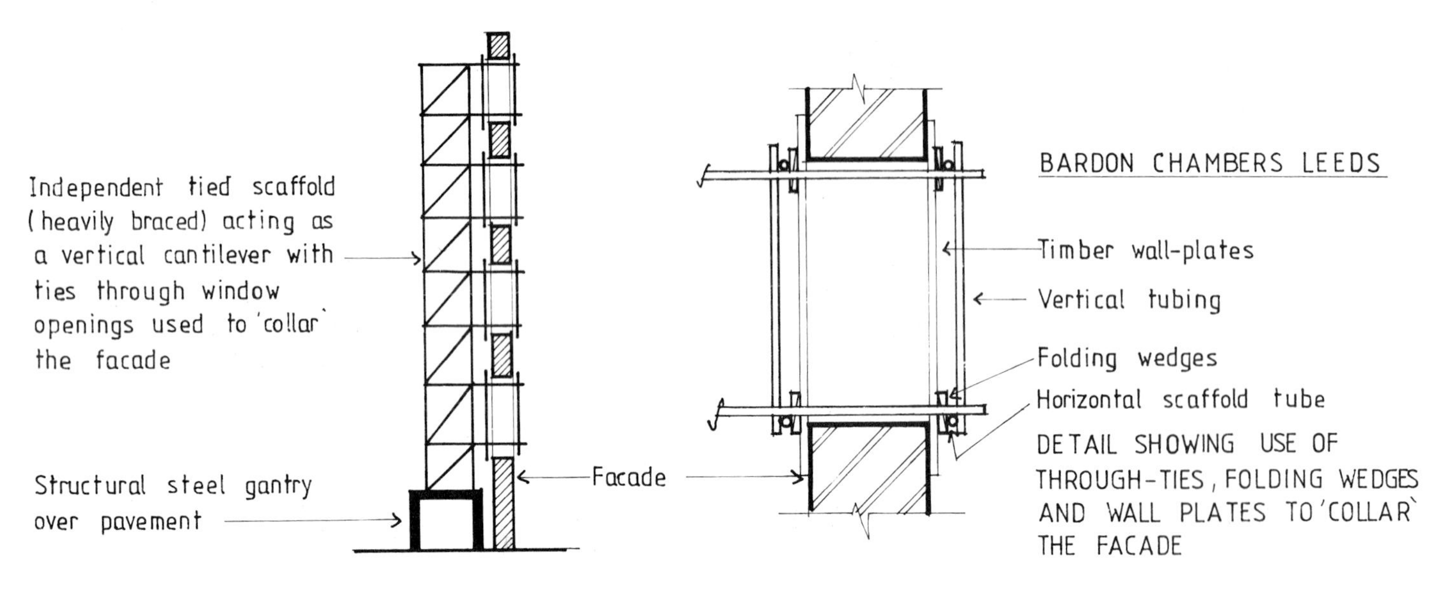

Figure 6.1 Typical example of wholly external temporary support system.

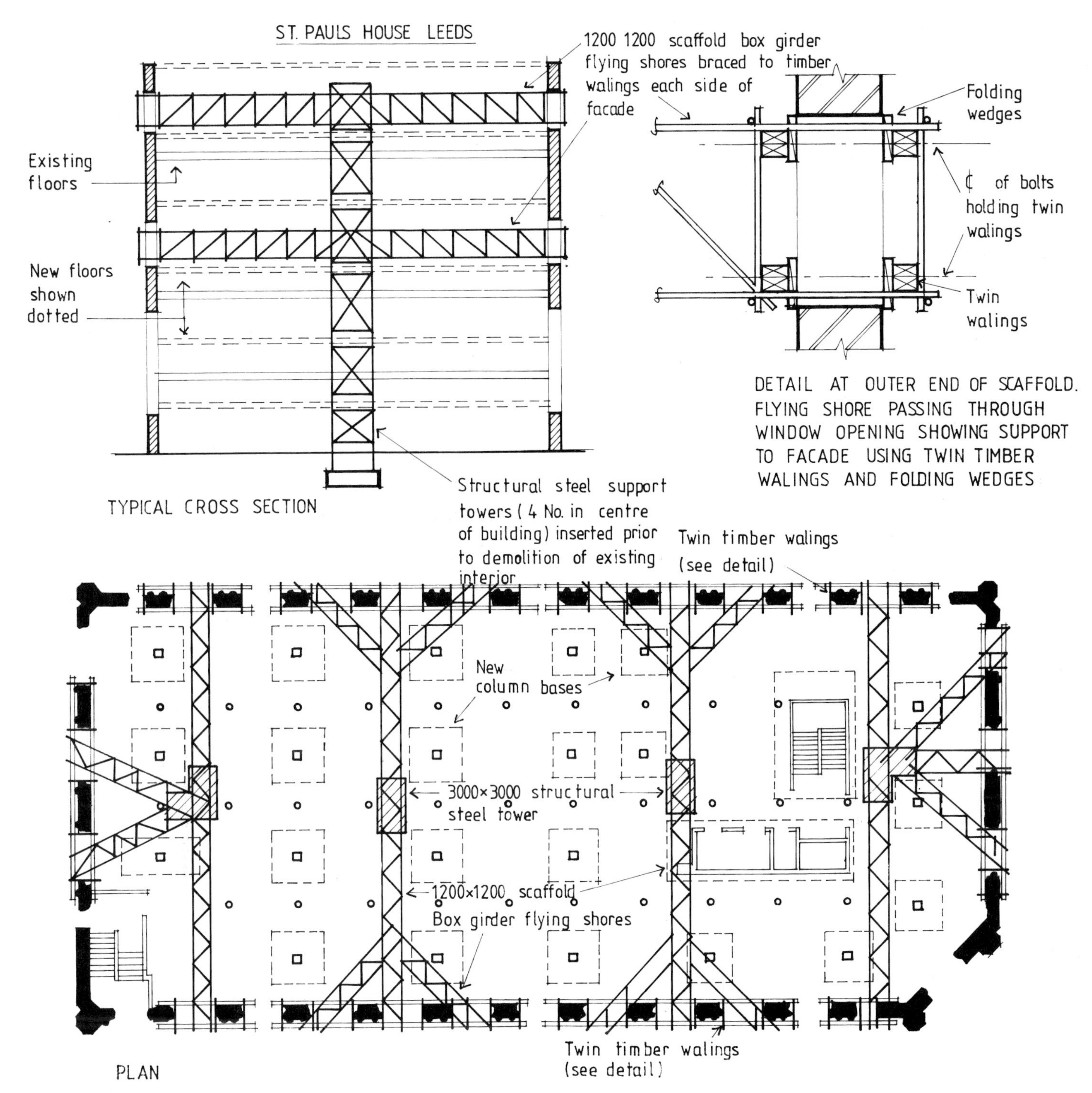

Figure 6.2 Typical example of wholly internal temporary support system.

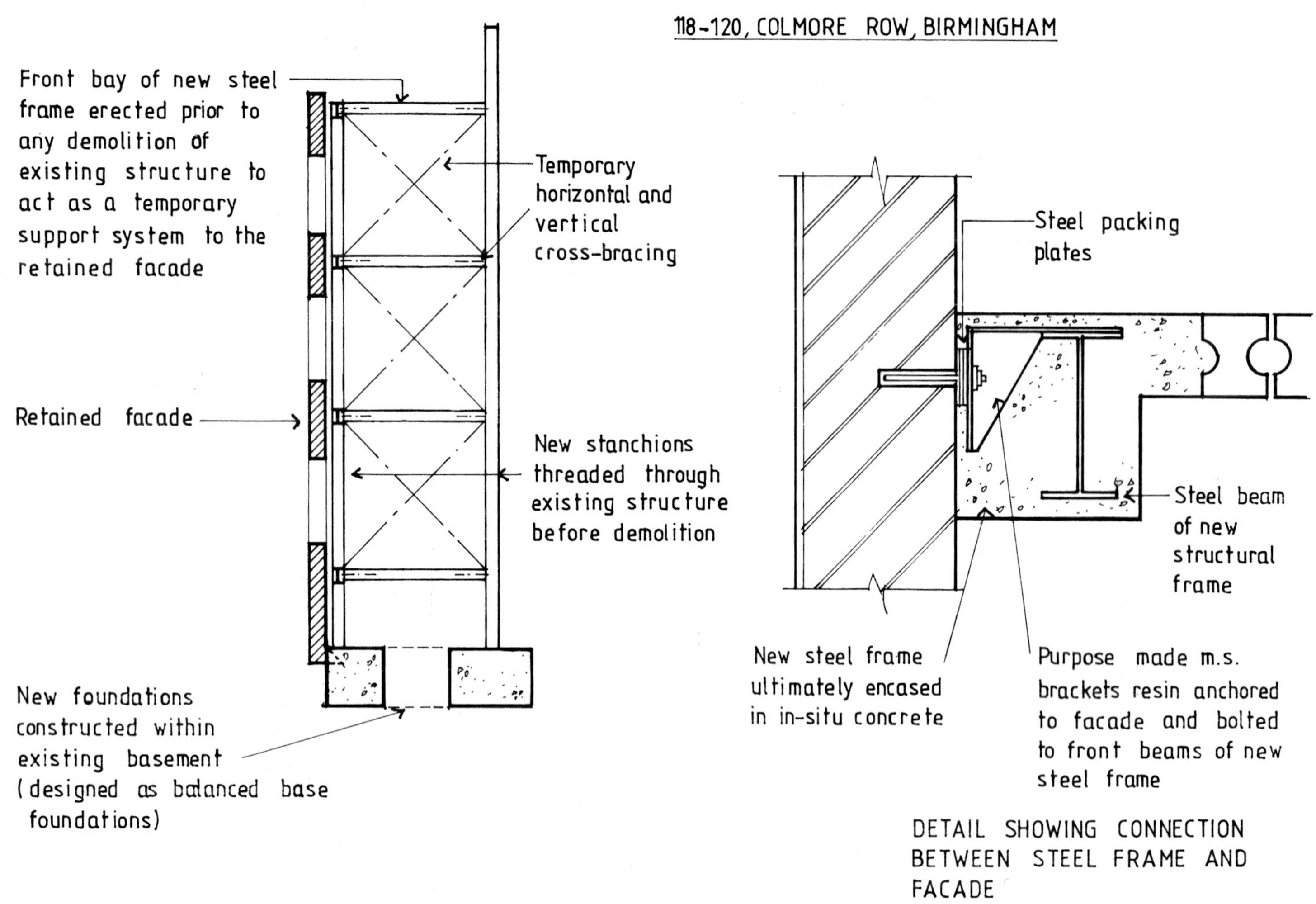

Figure 6.3 Internal facade support system employing part of new steel frame.

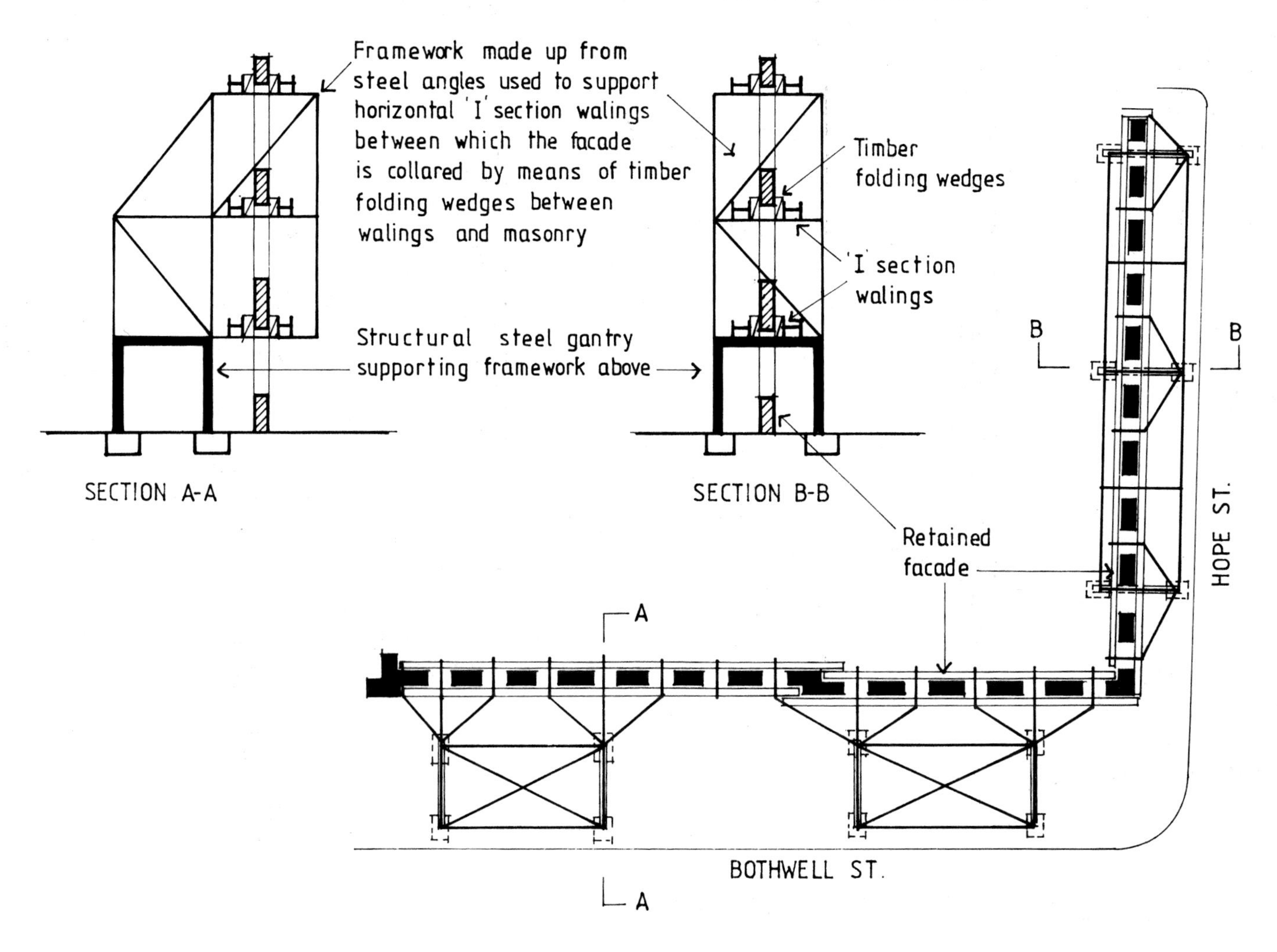

Figure 6.4 Typical example of part internal/part external temporary support system.

extensive period during the works whilst the building is 'opened-up' to the elements.

Generally, temporary support systems to retained facades fall into three categories: wholly external, located entirely outside the facade; wholly internal, located behind the facade within the zone of the existing and new structures; or part internal/part external, some of the supporting elements being located behind and some outside the facade. External support systems have the important advantages of not interfering with demolition or subsequent construction work, but they often obstruct adjacent footpaths and roadways, and in a number of cases have not been permitted for that reason. Internal support systems, on the other hand, leave adjacent thoroughfares unobstructed, but place severe constraints upon the progress and efficiency of both the demolition of the existing structure and the new construction work. Part internal/part external systems combine the advantages and disadvantages of both.

It is essential, whatever category of temporary support system is used, that it is installed, and capable of giving total support to the facade, before any demolition of the building's interior structure takes place, and that it remains in position until the facade is permanently tied back to the newly erected structure behind. It is clear, therefore, that if an internal support system is used, its design will be complicated by having to ensure that it does not conflict either with elements of the existing building or of the new building. In addition, some members of the internal support system will inevitably interfere with the demolition operations and the erection of the new structure. In certain cases, the nature of the scheme may require that the temporary support system is partly internal and partly external. The main difficulty with a part internal/part external support system is that it combines the disadvantages of both. Adjacent footpaths, and possibly roadways, may be obstructed by the external elements, and demolition and construction operations are interfered with by the internal elements.

Four examples of the various categories of temporary support system (three of which are described in greater detail in Chapter 7) are shown in Figures 6.1–6.4.

6.3 Facade ties

A technical problem common to all facade retention schemes involves devising a method of permanently tying back the retained facade to the new structure erected behind it. The lateral support formerly provided by the original internal structure must be replaced by some form of mechanical tie system between the facade and the new structure. These facade ties must fulfil a number of important functional requirements.

1. They must effectively hold back the facade and prevent any outward movement away from the new structure.
2. They must not transmit any vertical loads from the new structure to the facade (since the facade should not normally act as a load-bearing element in the new design)
3. They must be capable of accommodating any predicted differential settlement between the new structure and the retained facade without causing damage to the ties themselves, the facade or the new structure (see section 6.4).

The most widely used solution to this problem is to employ some form of resin anchor system which involves anchoring steel tie-bars into the facade masonry with a rapid-setting resinous mortar and then connecting them directly or indirectly to the new structure. The anchoring of the tie-bars into the facade masonry may be executed using either a resin cartridge method or pre-mixed resin. With the resin cartridge method, a preformed plastic or glass phial, containing the unmixed ingredients capable of forming the rapid-setting resinous mortar, is inserted into a pre-drilled hole in the masonry. The tie-bar is then spun into the hole using a drilling tool, breaking the cartridge and mixing its ingredients to form the mortar which anchors the bar firmly into the masonry. The alternative method is to pump pre-mixed resinous mortar into a pre-drilled hole and either spin or push the tie-bar into it. The resin anchoring procedure is illustrated in Figure 6.5 and typical cartridges and tie-bars are shown in Figure 6.6.

The connections between the projecting tie-bars and the new structure can be effected in a number of ways (Figure 6.7). Methods

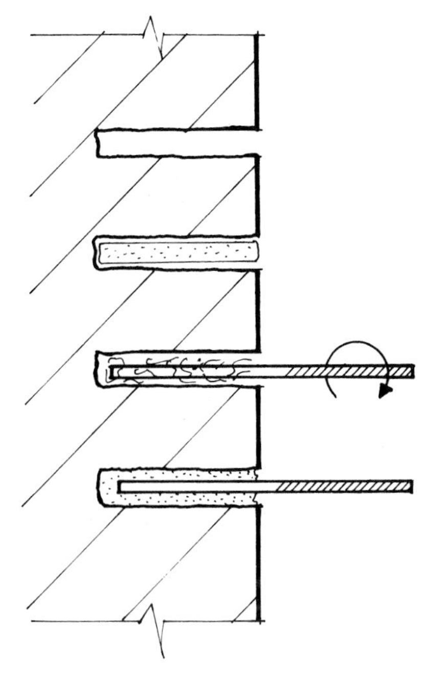

1. HOLE DRILLED INTO MASONRY

2. RESIN CARTRIDGE (containing unmixed ingredients required to form rapid-setting resinous mortar) INSERTED INTO HOLE

3. TIE BAR SPUN INTO HOLE WITH DRILLING TOOL (perforates cartridge and mixes ingredients)

4. TIE BAR ANCHORED INTO MASONRY BY RAPID SETTING RESINOUS MORTAR

* AS AN ALTERNATIVE TO RESIN CARTRIDGES PRE-MIXED RESINOUS MORTAR MAY BE PUMPED INTO HOLE AND TIE-BAR PUSHED IN

Figure 6.5 Resin-anchoring technique for installing facade-ties.

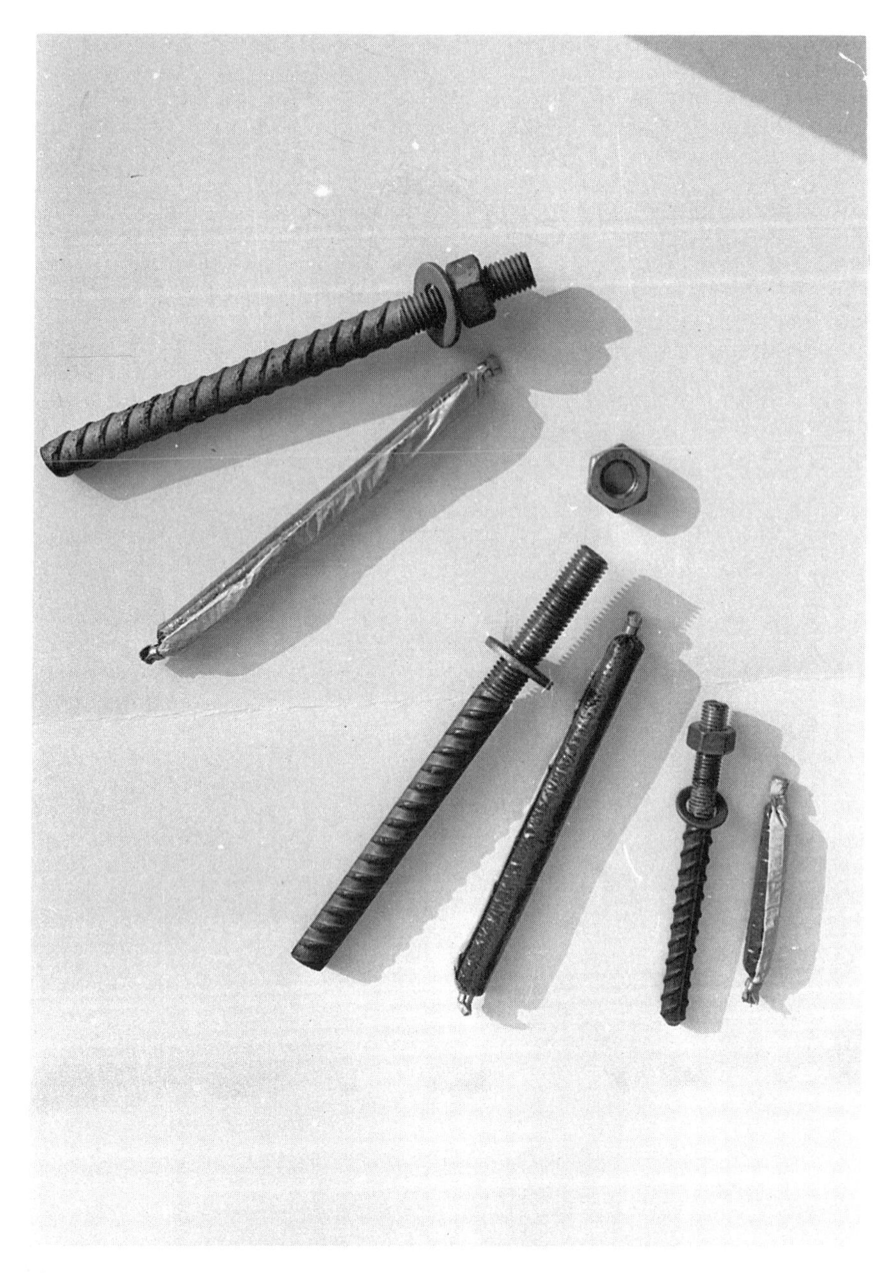

Figure 6.6 Resin-anchor cartridges and tie bars.

used include casting the projecting tie-bars directly into the edges of the new floor slabs, or by using indirect connections where steel angles, bolted to the edges of the new structure, are fastened to the projecting resin-anchored tie-bars using locknuts.

As an alternative to using resin-anchored tie-bars, various forms of through-tie may be used. These comprise steel bars passing completely through the facade, and secured to steel plates on the external face. The inner ends of the tie-bars project from the internal face of the facade and are secured to the new structure directly or indirectly in the same way as resin anchored tie-bars. One problem with this through-tie method involves the concealment of the outer ends of the tie-bars and anchor plates, which is relatively easy with stuccoed or rendered facades but more difficult with masonry or brickwork.

A typical example of a through-tie is shown in detail A, Figure 7.7, page 54.

6.4 Differential settlement

It is essential in facade retention schemes that the detailing at the junction of the new structure and existing facade allows settlement of the former to occur if this has been predicted. A rigid connection between the new structure and the facade could result in potentially serious structural damage in the event of settlement. If such settlement has been predicted, its extent should be calculated using data from the subsoil investigations. The facade ties and interface detail between the facade and the new structure should then be designed to allow this settlement to take place without causing damage to either of the structures or the facade-ties themselves. One of the most effective ways of achieving this is to use an indirect steel angle tie with a vertical slotted hole through which the resin anchored tie-bar passes (Figure 6.8). The slotted hole in the vertical leg of the angle enables the new structure to settle without causing damage.

The interface treatment between the new structure and the retained facade must also be considered where settlement of the former has been predicted. The interface between the new structure and the retained facade is usually at the outer faces of the new structure's

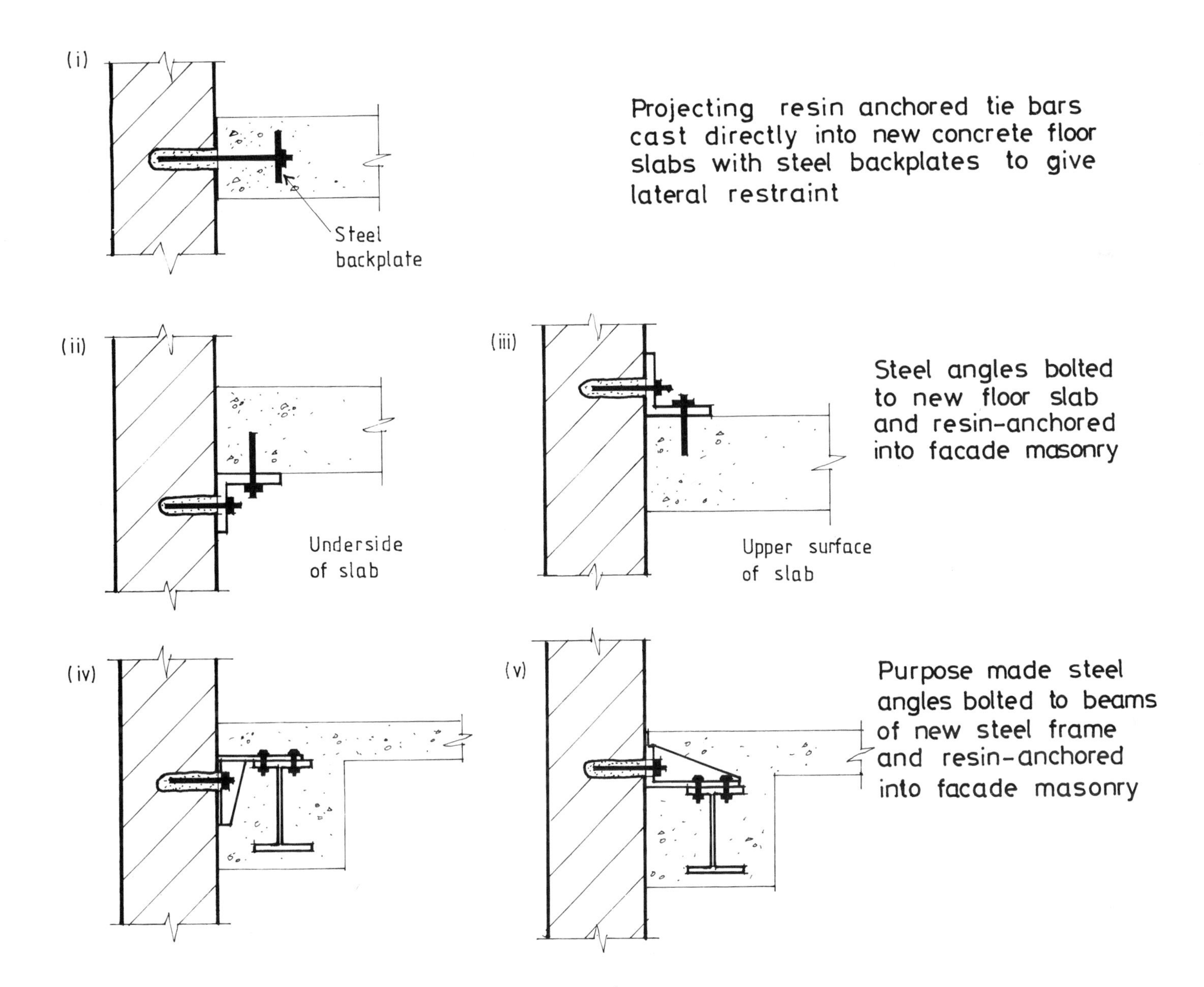

Figure 6.7 Various methods of connecting resin-anchors to the new structure.

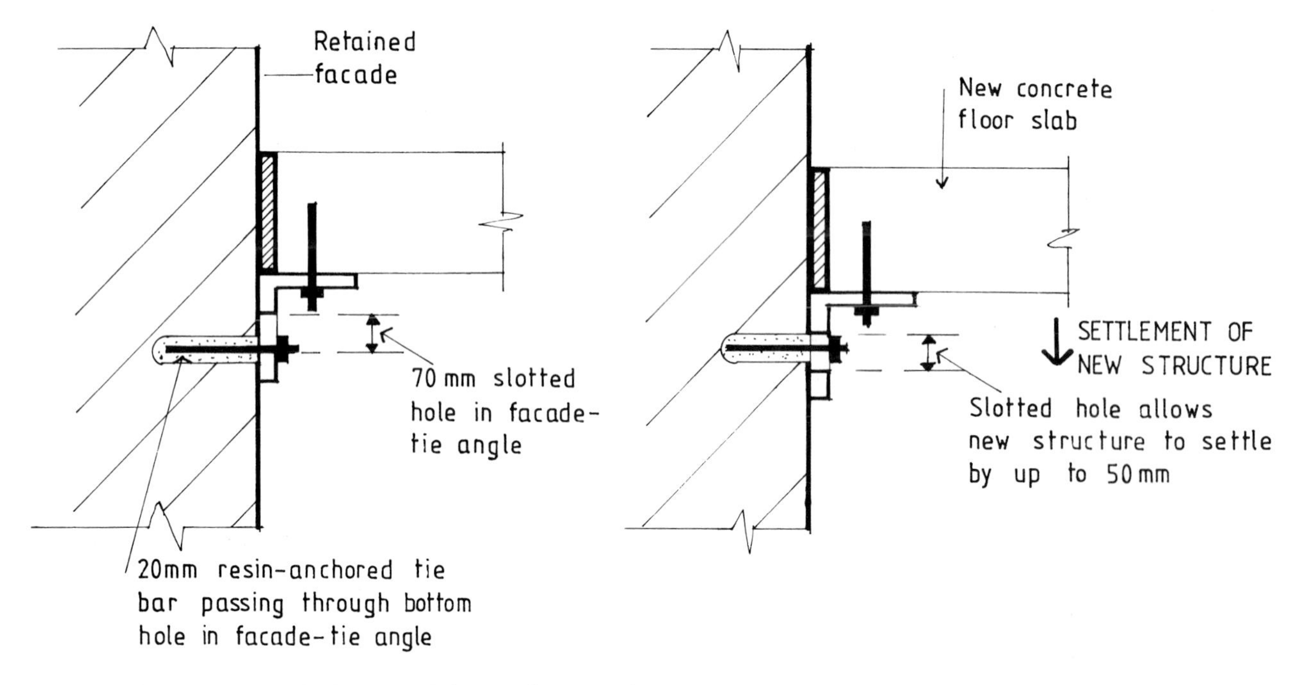

Figure 6.8 Facade-tie design permitting settlement of new structure.

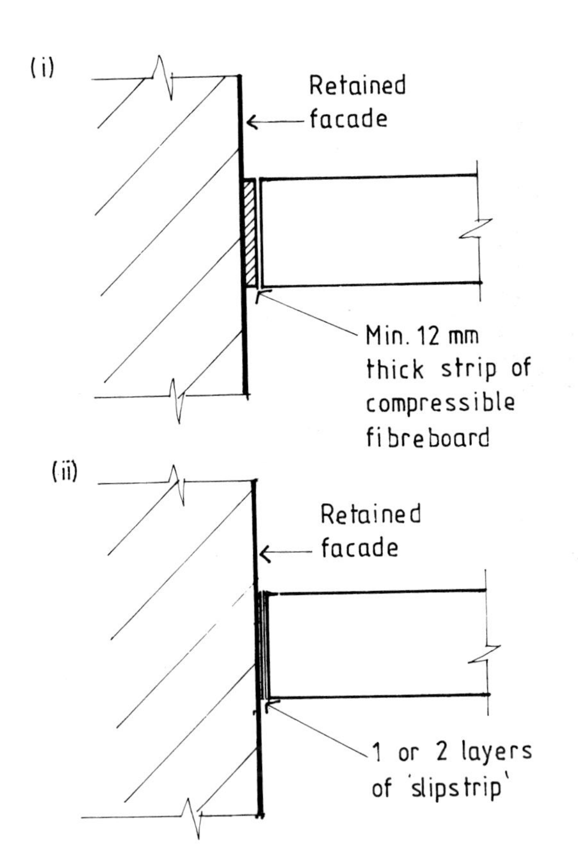

Strip of compressible fibreboard between new floor slab edges and inside face of facade

One or two layers of 'slipstrip' (e.g dense polythene) between new floor slab edges and inside face of facade

Figure 6.9 Interface details permitting settlement of new structure.

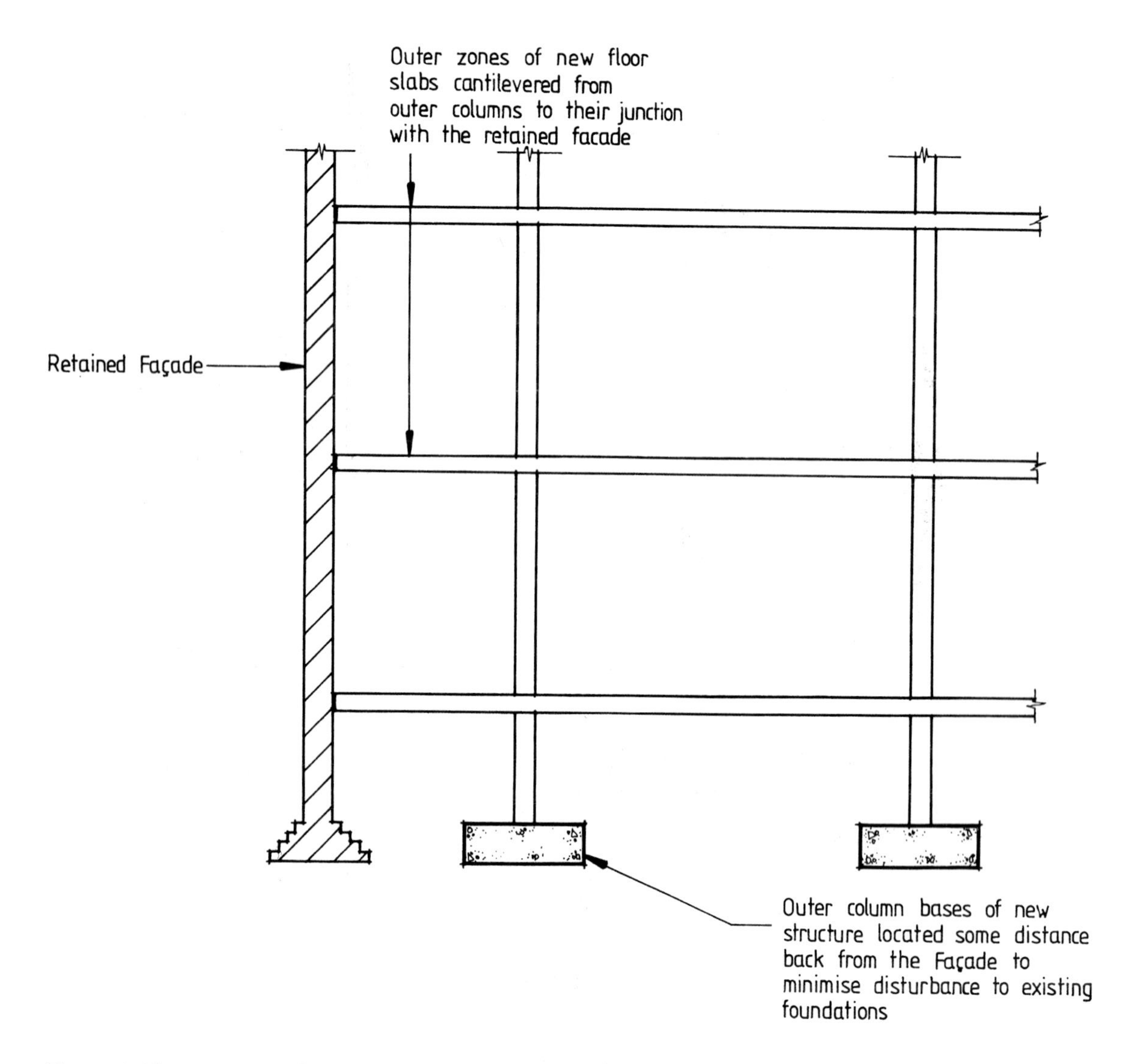

Figure 6.10 Foundation design to prevent damage to the stability of the retained facade.

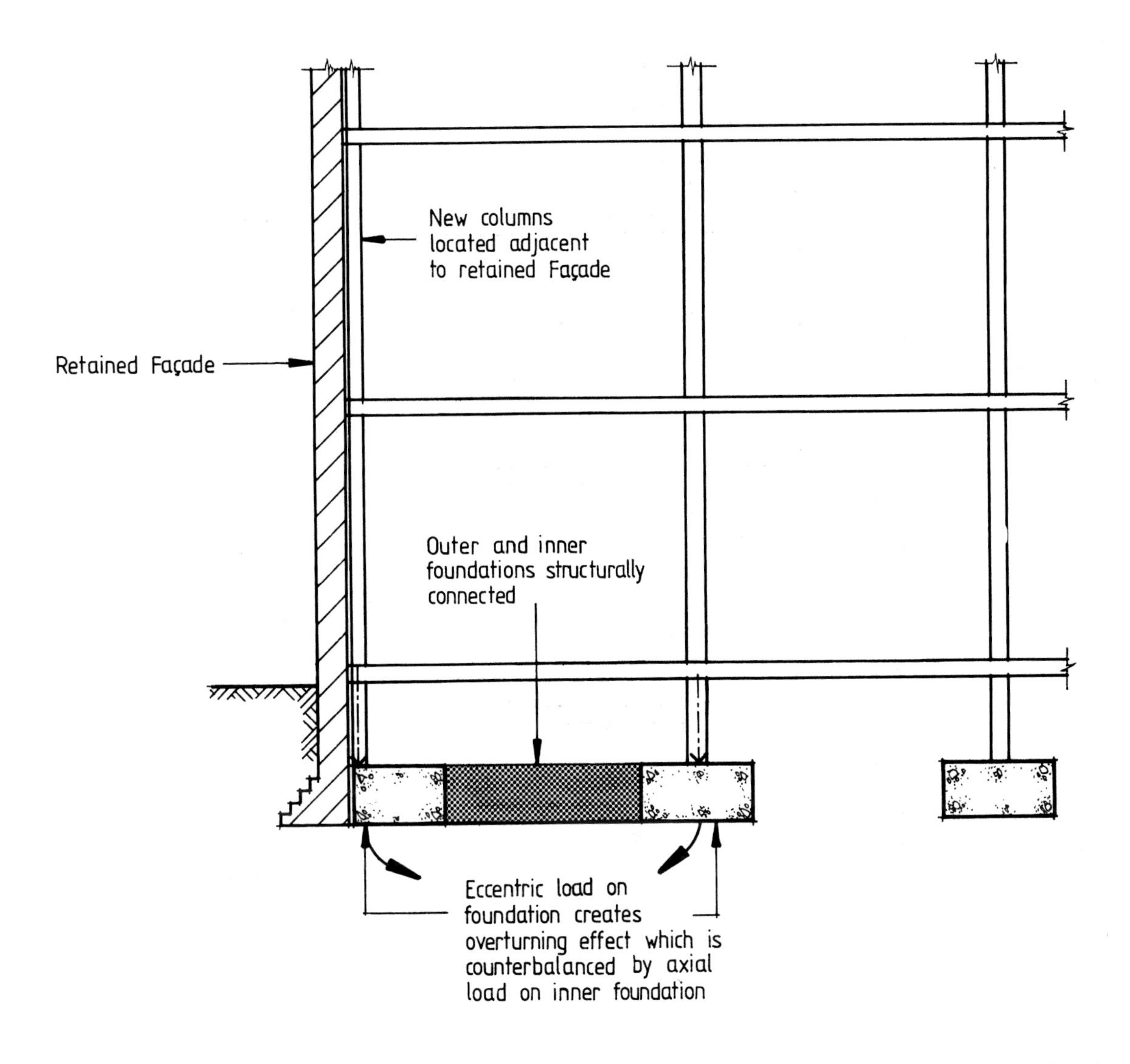

Figure 6.11 Balanced-base foundations used to prevent damage to the stability of the retained facade.

columns and/or the edges of its floor slabs. The most common method of allowing for differential movement here is to provide some form of slip surface between the new and existing elements which will prevent bonding of the two surfaces. The slip surface may comprise single or multiple layers of dense polythene or similar material, or a thin layer of fibre-board. Another accepted interface treatment is to leave a narrow gap where the new structure meets the facade to ensure that settlement may take place without damage. Typical interface treatments are shown in Figure 6.9.

6.5 Foundation design

It is essential that the design and construction of the new structure does not adversely affect the stability of the retained facade, and this is of particular importance at foundation level. The foundations to many historic buildings are often found to be weak and unstable and are therefore vulnerable to any disturbance caused by new construction works. The two most common solutions used to overcome this problem are as follows.

The first, and probably the simplest method is to locate the outer column bases of the new structure some distance back from the retained facade and to cantilever the new floors from these columns to their junction with the facade (Figure 6.10). This ensures that no new foundations are constructed adjacent to those of the facade, therefore ruling out any constructional disturbance and subsequent harmful effects caused by the new structure's loads.

The second method, illustrated in Figure 6.11, is employed when the design of the new structure requires that some columns must be located immediately adjacent to the retained facade. In order to minimize disturbance at the base of the facade, the new column bases are constructed immediately adjacent to it, but do not undermine it. The new columns, which are also immediately adjacent to the facade, inevitably subject these bases to eccentric loading and an overturning effect, which must be counteracted in some way if the foundations are not to fail. The eccentric loads are 'balanced' by structurally connecting these bases to the axially loaded bases of an inner line of columns. This use of 'balanced-base' foundations counteracts the overturning effect which the columns adjacent to the facade have on their bases.

In certain facade retention schemes, some undermining and underpinning of the existing foundations is inevitable. In these cases it is of paramount importance that the new foundations are designed, and their construction phased, so as to minimize any adverse effects they may have on the stability of the retained facade.

Further reading

Highfield, D. (1982) *The Construction of New Buildings Behind Historic Facades: The Technical and Philosophical Implications*, M. Phil. thesis, University of York, York.

Highfield, D. (1983) Keeping up Facades, *Building*, **245**(39), 40–1.

Highfield, D. (1984) Building Behind Historic Facades, *Building Technology and Management*, **22**(1), 18–25.

Highfield, D. (1987) *Rehabilitation and Re-use of Old Buildings*, E. & F.N. Spon (An imprint of Chapman and Hall), London.

7
Facade retention case studies

The wide variety of buildings that have been the subjects of facade retention schemes has resulted in a range of different solutions to the associated technical problems, and these have been described in the previous chapter. The aim of this chapter is to give a detailed insight into the means used to overcome the technical problems for a number of successfully completed schemes. Eight case studies have been selected for inclusion in order to cover the methods most widely used to overcome the problems of providing temporary support, tying back the facade to the new structure, differential settlement and foundation design. Background information on each case study is given, followed by a comprehensive description of all the relevant technical aspects, supplemented by detailed drawings and photographs.

Case study one

118 – 120 Colmore Row, Birmingham

Background to the scheme

Although the existing building was not listed, its facade comprised an attractive example of late Victorian architecture, blending well, in both scale and character, with the surrounding buildings on Colmore Row, within one of the city's outstanding conservation areas. For this reason, the local planning authority recommended that any redevelopment of the building should at least involve preservation of its facade. The existing interior proved unsuitable for conversion to provide the lettable office accommodation required, and was therefore demolished and replaced by a new steel framed structure erected behind the existing facade which was retained in accordance with the planners' wishes. The existing building comprised three storeys and a basement, and in common with most facade retention schemes, the replacement structure provided additional floor levels. This was achieved by providing a stepped mansard roof to the new structure, which enabled two additional floors to be inserted above the original roof level.

The new steel framed structure, which was encased in *in-situ* concrete, supported pre-stressed concrete rib and filler block floors with an *in-situ* structural concrete topping. Externally, the stepped mansard roof was finished in natural slate and the stuccoed facade repaired and re-decorated. Figure 7.1 shows the new front elevation, and Figures 7.9 and 7.10 show external views of the completed scheme.

Temporary support system

The temporary support system to the retained facade was wholly internal, as a result of the local authority's stipulation that the adjacent pavement should not be blocked by external supports. Rather than employing a truly temporary support system, the structural frame to the new building was designed so that part of it could be initially employed as a support to the retained facade during the works. The first bay of the new structural steel frame, immediately behind the facade, was erected and the facade tied to it before any major demolition took place. The erection of this new steelwork within the confines of the existing structure was one of the most technically complex phases of the whole project. The new balanced-base foundations had to be constructed within the extremely restricted conditions of the existing basement, and selective demolition of sections of walls and floors was required to allow the eight stanchions, the beams, and the temporary bracings to be threaded into the existing structure and fixed into their respective positions (Figure 7.11).

The only truly temporary elements of the support system were the horizontal and vertical cross-bracings installed to increase its stability and ensure the provision of adequate support to the facade. All of the other members of the support system comprised permanent elements of the new steel frame, therefore effecting considerable cost savings over the more conventional temporary support systems used on other case studies. This use of part of the new permanent structure to provide temporary support to the retained facade gave considerable savings in the time, labour and materials which would have been necessary to erect, and subsequently dismantle, a truly temporary support system. An additional advantage was that the support system comprised far fewer elements than a typical scaffolding system, therefore interfering to a lesser extent with both demolition work and construction operations.

On completion of the structural steel supporting bay, the facade was tied back to it by means of specially designed resin-anchor ties which would also subsequently act as the permanent facade ties. These resin-anchor ties comprised purpose-made mild steel brackets fixed to the beams of the supporting structure by bolting to their bottom flanges, and fixed to the facade using resin-anchor bolts. The ties were fixed in temporary positions whilst they served as elements of the temporary support system, and later adjusted to their final positions to act as permanent facade ties holding the facade back to the new structure. The facade ties, in both their temporary and permanent conditions, are fully described later. In addition to the resin-anchor ties, a number of through-ties were employed at the uppermost level, passing from the beams of the supporting structure, through the full thickness of the facade and fixed to plates on its outer face.

After the installation of the resin-anchor ties and the through-ties, the facade was effectively secured to its temporary support system and demolition of the existing internal structure could commence. Figure 7.5 shows the layout of the temporary support system and Figures 7.11 and 7.12 show parts of the system before and during demolition of the existing building.

Facade ties

The facade was initially tied back to the temporary support system, and finally to the new steel frame, by means of resin-anchor ties at three levels, together with a number of through-ties at the uppermost level. As previously explained, the first bay of the new steel frame had to be erected prior to any demolition of the existing structure to act as the facade's temporary support system. In addition to the problems of installing the temporary support system within the existing building, the sequence of operations was further complicated by the fact that the floor levels of the existing building conflicted with those of the new building. This problem was overcome by locating the new steel floor beams of the first bay in temporary positions above the existing floors (see Figure 7.6). When the first bay of the new steel frame, with its beams in their temporary positions, had been erected and braced, the installation of the facade ties in their initial temporary support positions was commenced.

The resin-anchor ties between the front beams of the structural steel frame and the facade masonry are detailed in Figures 7.6 and 7.7, and comprised purpose-made mild steel brackets, bolted initially to the bottom flanges of the beams and resin-anchored to the facade masonry. The mild steel brackets were fixed to the bottom flanges of the supporting beams using two 20 mm diameter mild steel bolts, and fixed to the facade using a single stainless steel stud resin-anchored into the masonry. 'Kemfix' resin-anchors were used for the facade fixings, each connection being made as follows.

1. 24 mm diameter × 145 mm deep hole drilled into masonry using rotary percussion drill, and all loose dust cleaned out.
2. 'Kemfix' resin capsule, size M.20, inserted into hole. The 'Kemfix' resin capsule comprises a sealed glass tube containing polyester resin, quartz granules and a phial of hardener which, when the capsule is broken and they are mixed, form a rapid-setting resinous mortar.
3. 20 mm diameter × 240 mm long threaded stainless steel stud spun into the hole using a rotary percussion drill. This operation breaks the 'Kemfix' capsule, mixes the ingredients and embeds the stud in the resulting resinous mortar.
4. Mild steel facade-tie bracket secured to projecting threaded end of stainless steel stud using nut and washer.

Each mild steel facade-tie bracket was provided with two 21 mm diameter holes in its vertical leg through which the projecting resin-anchored stud passed in order to secure it to the facade. The lower hole was used to locate the bracket in its temporary position, and the upper hole to allow re-connection of the bracket in its final, permanent position (see Figure 7.6). To allow for variations in the gap size between the mild steel brackets and the uneven facade, thin mild steel packing plates were used as required to fill the space between the vertical legs of the brackets and the inside face of the facade.

A total of 51 resin-anchor connections were used to tie back the facade, initially to the temporary support system, and finally to the new structural frame. 21 were located at 750 mm centres at first floor level, 21 at 750 mm centres at second floor level and 9 at third floor level. Those at the upper level were used in conjunction with a number of through-ties of different design which are described later. Figure 7.1 shows the locations of the facade ties.

When the 51 resin-anchor ties between the facade and its temporary support system were completed, demolition of the existing structure could commence.

After the demolition had been completed, the front beams, to which the facade was tied, could be lowered from their temporary positions into their permanent positions and the facade re-connected to them. This operation was carefully phased, only one beam being re-positioned at a time, therefore ensuring that the facade's temporary support was only partially reduced each time a beam was lowered. The design of the mild steel facade-tie brackets and the locations of the resin-anchored studs was such that no further resin-anchoring into the masonry was necessary in order to re-connect the facade to the supporting beams after they had been lowered into their permanent positions. The procedure for re-positioning the beams and permanently tying the facade back to them was as follows (read in conjunction with Figure 7.6).

1. Mild steel facade-tie brackets along the beam to be lowered removed by disconnecting them from bottom flange of beam and from projecting resin-anchored studs.
2. Beam disconnected from stanchions and lowered from its temporary position into its permanent position.
3. Mild steel facade-tie brackets re-connected to projecting resin-anchored studs (using upper holes in vertical legs of brackets), and to top flange of lowered beam.

Figure 7.13 shows a beam in its permanent position, with the facade-tie brackets fixed to its top flange.

After every beam had been re-positioned in this manner, and the facade-tie brackets re-connected, the 51 resin-anchor facade-ties were complete. The final operation, required to complete the tying back of the facade to the new structural frame, was the insertion of 12 through-ties at third floor level. Each of these through-ties (Type A in Figure 7.1, and detail A in Figure 7.7) comprised a 20 mm diameter stainless steel bolt passing through the full thickness of the wall and connected at its inner end to the steel supporting beam, and at its outer end to the outer face of the facade. The inner end connection was effected by passing the bolt through a hole in the web of the supporting beam and securing it with a nut and washer. The outer end connection, designed to prevent the facade moving outwards, away from the new structure, comprised a 250 × 250 × 10 mm thick stainless steel plate fastened into the outer face of the wall using a nut and washer. The plate was set into a 50 mm deep recess in the facade's outer face which was subsequently filled and decorated to match the existing stucco in order to conceal the plate and fixing.

When all of the facade-ties were completed, the remainder of the new steel frame was erected and the cross-bracings from the front bay removed. The steel frame was then encased in *in situ* concrete which also enveloped the facade-ties, providing permanent protection against fire and corrosion (Figure 7.7).

Differential settlement

The designs of the facade-ties and the detailing at the junction of the new structure with the retained facade included no provision for differential settlement. Both types of facade-tie comprised fully rigid connections to the new structure and to the facade, and no slip surface was provided at the junction of the concrete-encased frame with the facade (Figure 7.7). The sub-soil investigations concluded that the only likely settlement of the new building would be immediate settlement which would be complete before the *in situ* concrete casing was provided to the steel frame. It was this concrete casing around the front beams and columns that formed the interface between the new structure and existing facade and, because negligible settlement was expected after provision of the casing, no slip surface was provided. If, on the other hand, further settlement had been expected, it would have been necessary to provide a slip-surface between the new structure and the existing facade to prevent their bonding, thereby ruling out the possibility of any structural damage occurring as settlement took place.

The anticipated magnitude of the new structure's immediate settlement was estimated as only 10 mm, some of which would occur after the anchoring of the facade-ties into the masonry. It was predicted that such minimal settlement would result in minor local crushing of the facade masonry, caused by a slight bending of the resin-anchored tie-bars as the settlement took place (Figure 7.8). Both the bending of the tie-bars and the crushing of the masonry were considered to present a negligible risk of damage to the facade-ties or to the new and existing structures and, for this reason, no provision for differential settlement was necessary in the design of the facade ties.

Foundation design

The design of the new structural frame, and the fact that part of it was used as the temporary support system, required some columns to be located immediately adjacent to the facade. In addition, any undermining of the existing facade beyond its outer face was considered undesirable and this inevitably resulted in the front columns applying eccentric loads to their new foundations (Figures 7.3 and 7.5).

This problem was overcome by incorporating these foundations into a balanced-base arrangement, and structurally connecting them to the axially loaded bases of an inner line of columns in order to counterbalance the overturning moments created by their eccentric loads, as described in section 6.1.4. The resulting foundation arrangement is shown in Figure 7.3.

Architects Rolfe Judd Group Practice, London

Consulting engineers Chamberlain and Partners, Brentwood, Essex. Roy Bolsover and Partners, Birmingham.

Main contractor J.S. Bloor (Construction) Ltd, Tamworth, Staffordshire.

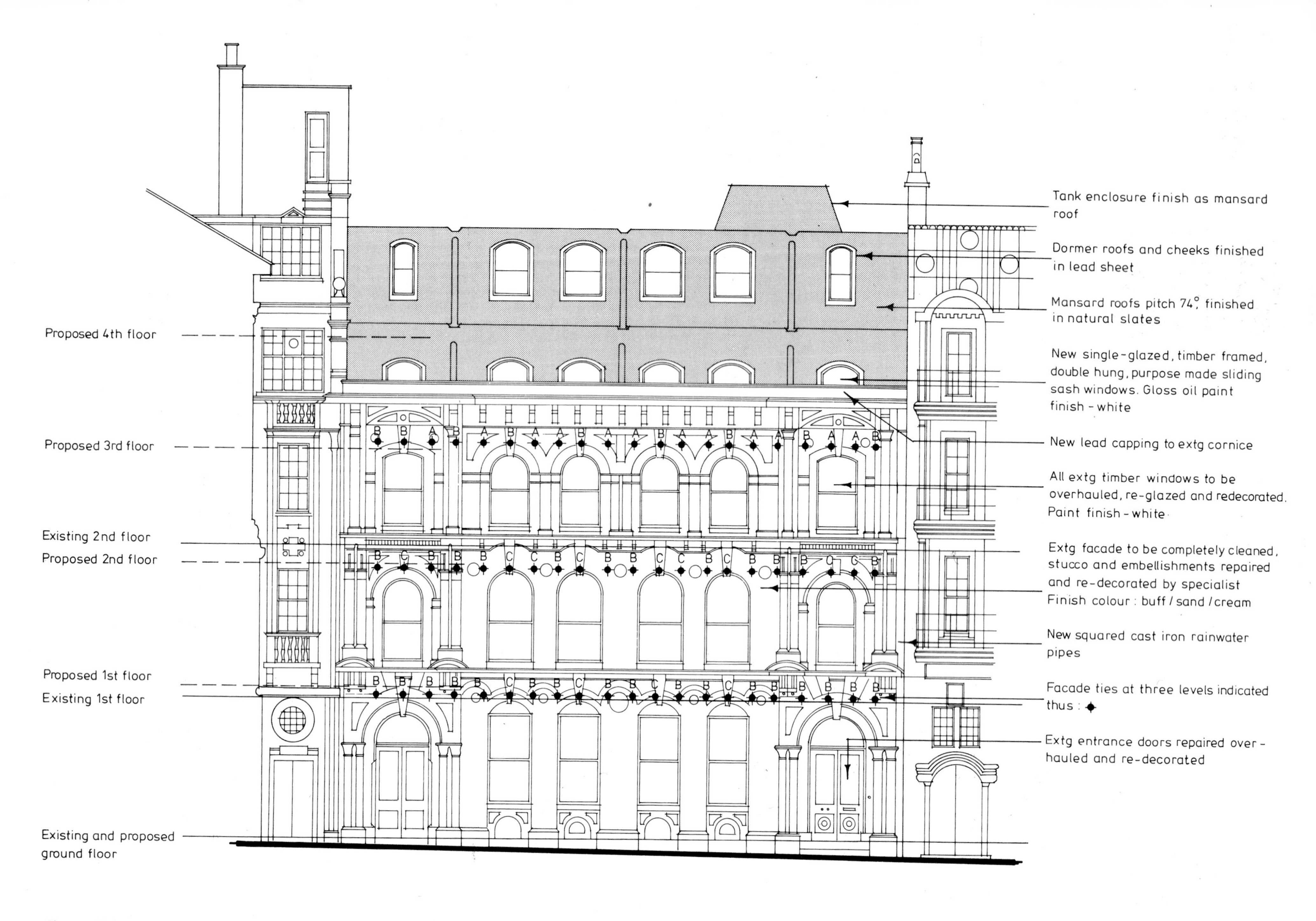

Figure 7.1 118–120 Colmore Row, Birmingham: proposed front elevation showing locations of facade-ties.

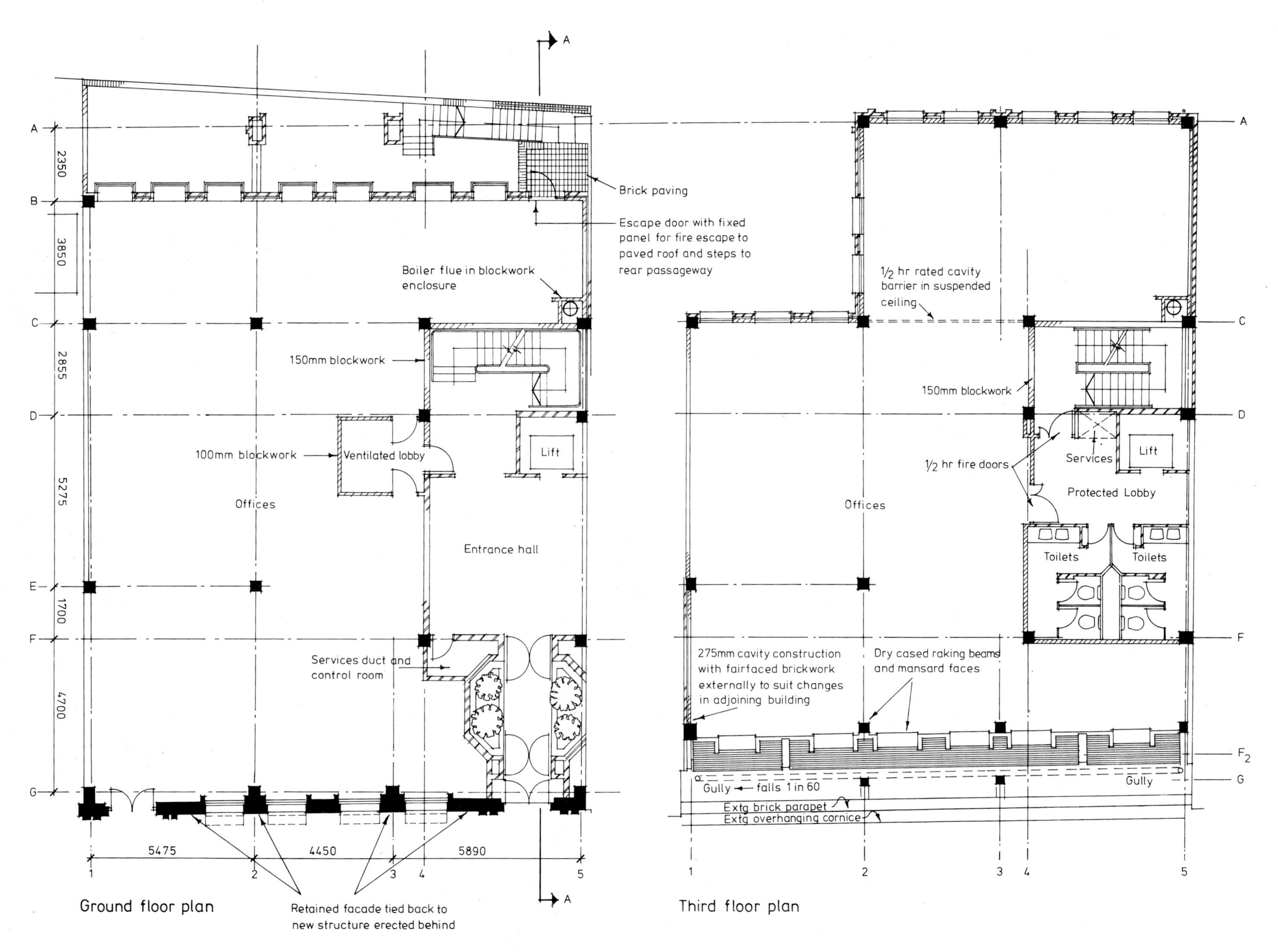

Figure 7.2 118–120 Colmore Row, Birmingham: ground and third floor plans.

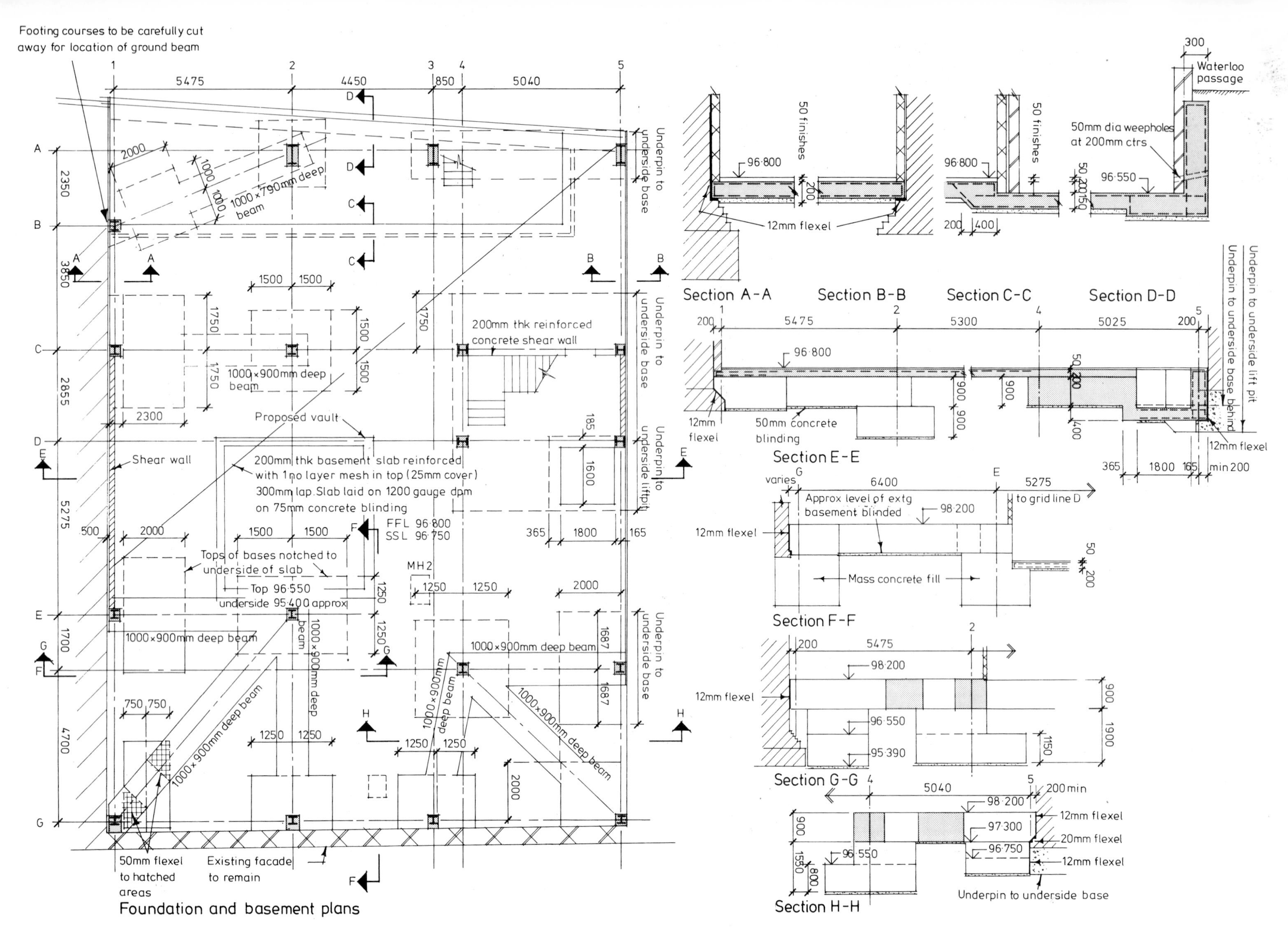

Figure 7.3 118–120 Colmore Row, Birmingham: foundation and basement plans and sections.

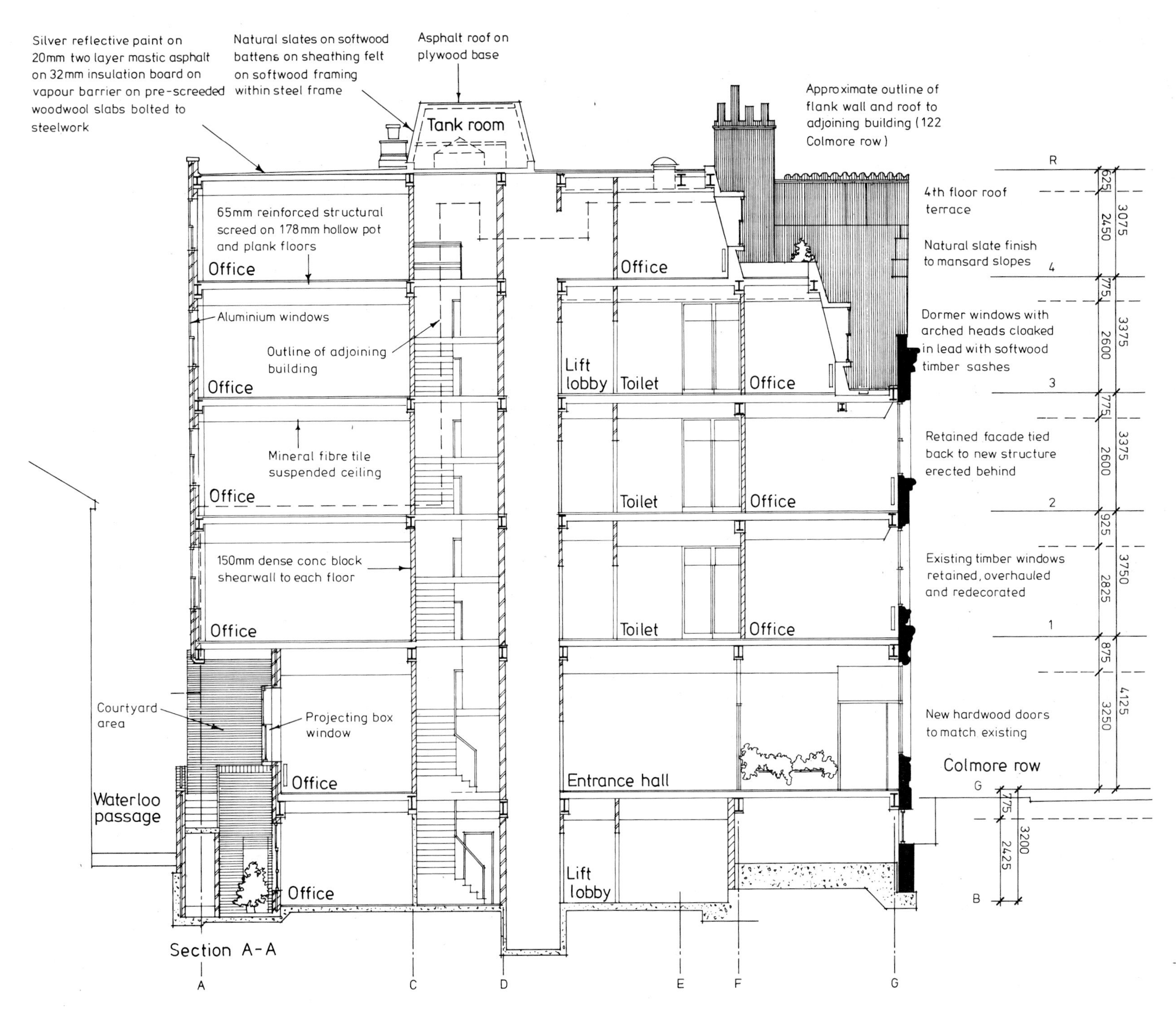

Figure 7.4 118–120 Colmore Row, Birmingham: vertical cross section.

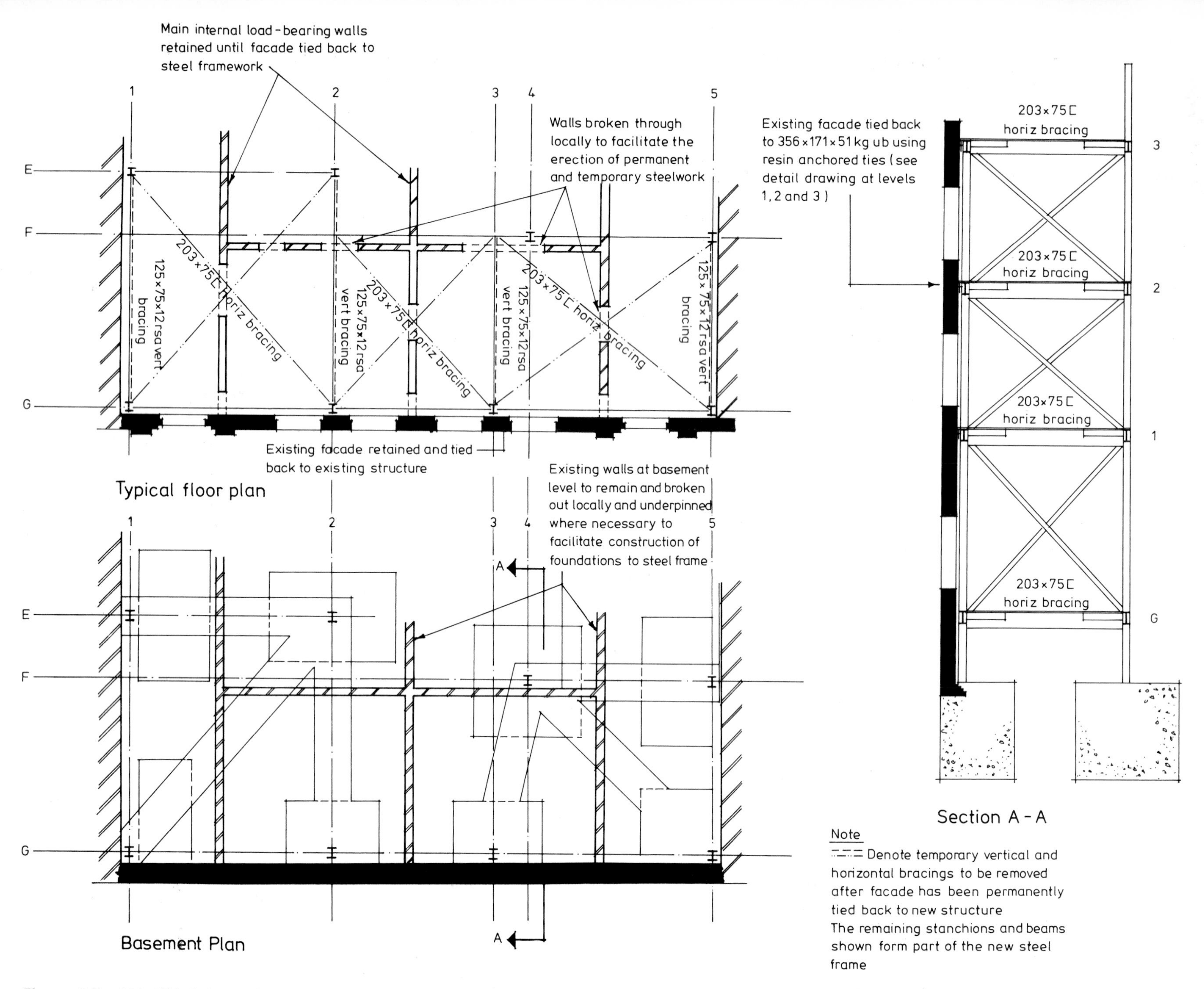

Figure 7.5 118–120 Colmore Row, Birmingham: temporary support system.

Facade ties in temporary position prior to demolition

G

175 varies

Beam in temporary position (prior to demolition of extg structure) and connected to ms bracket via lower flange

Ms purpose made brackets at 750mm ctrs

2 no 20mm dia bolts

'Kemfix' chemical anchor

Kemfix chemical anchor M20 with 20mm dia stainless steel stud washer and nut. Hole in wall to be 24mm.
Note – if when fixing the anchors the resin leeches into voids in the masonry another capsule is to be used.

Final position of beam

Extg floor joists preventing permanent positioning of new beams until after demolition

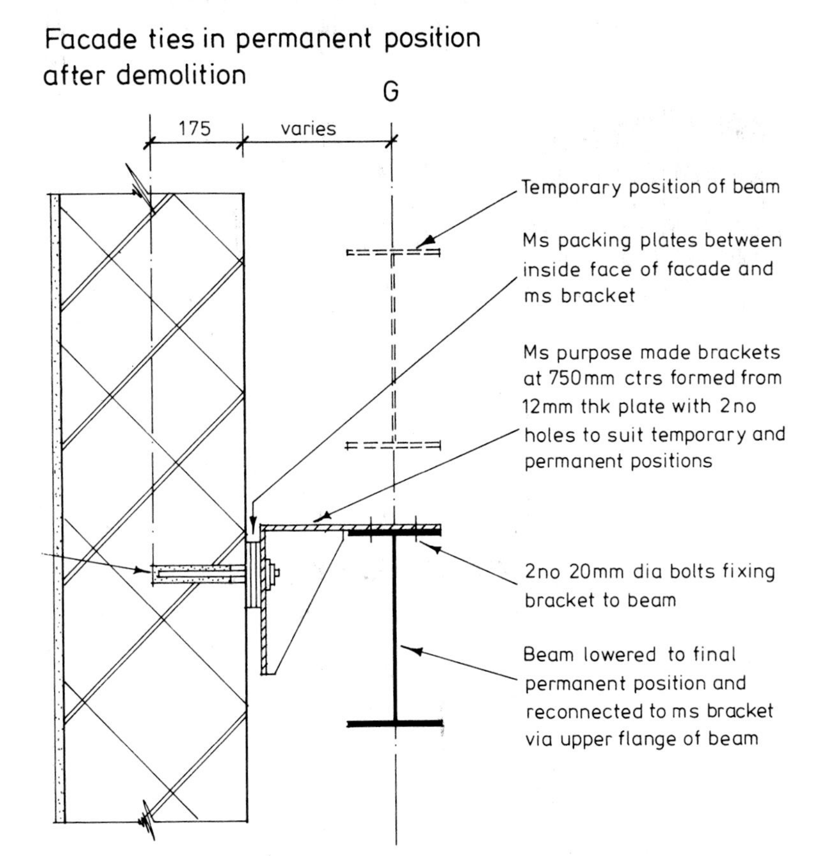

1 Beams of new steel frame located in temporary positions prior to demolition of extg internal structure (temporary positioning necessary due to positions of extg floor joists)

2 Purpose made ms brackets resin anchored to facade at 750mm ctrs and bolted to lower flanges of beams prior to demolition of extg internal structure to effect support to retained facade

3 Existing internal structure demolished

4 Purpose made ms brackets disconnected from lower flanges of steel beams and from resin anchored studs

5 Beams of new steel frame lowered into permanent positions

6 Ms brackets re-connected to upper flanges of steel beams and to resin anchored studs to effect permanent support to retained facade

Figure 7.6 118–120 Colmore Row, Birmingham: temporary support system/facade-ties.

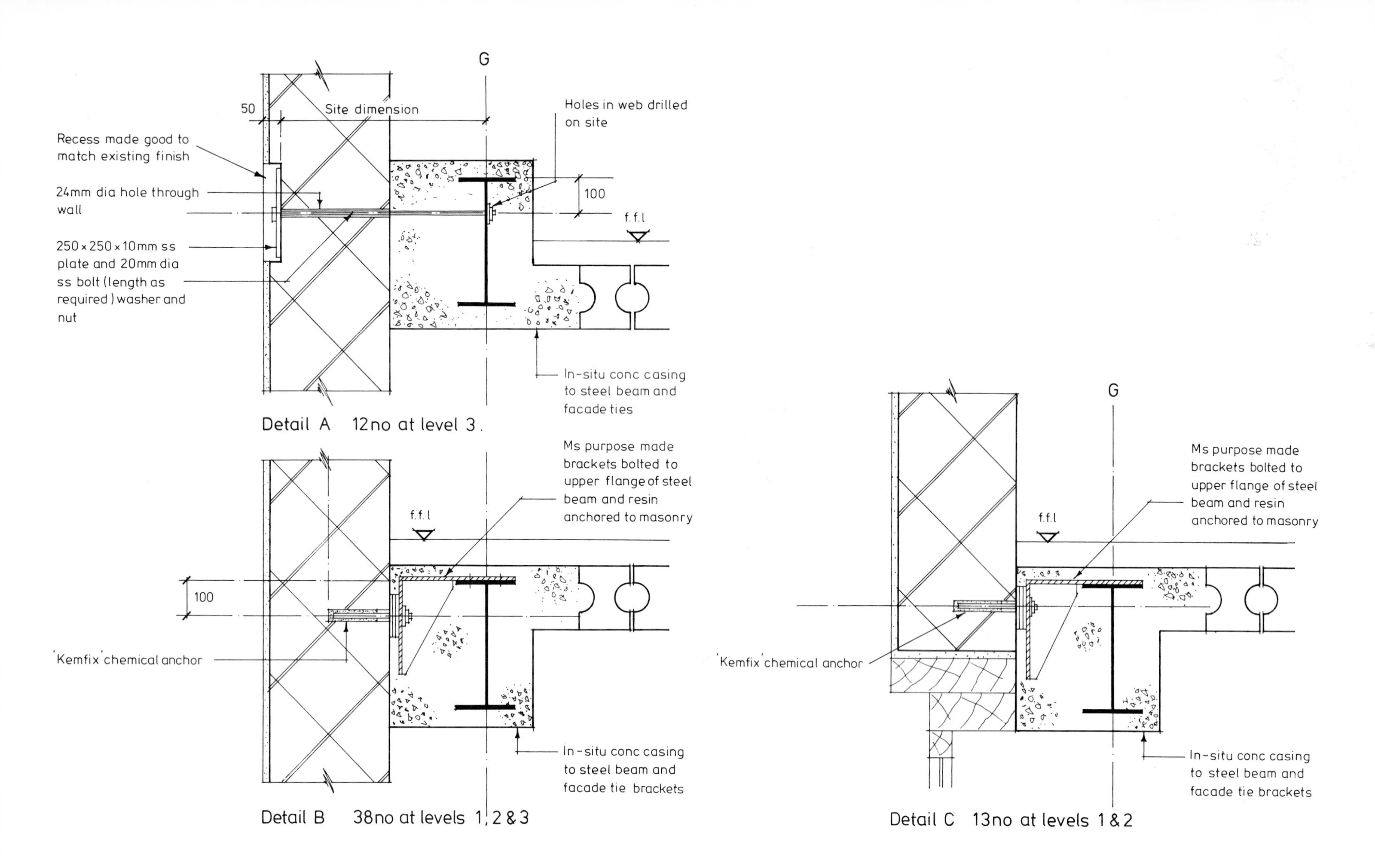

Figure 7.7 *118–120 Colmore Row, Birmingham: facade-ties.*

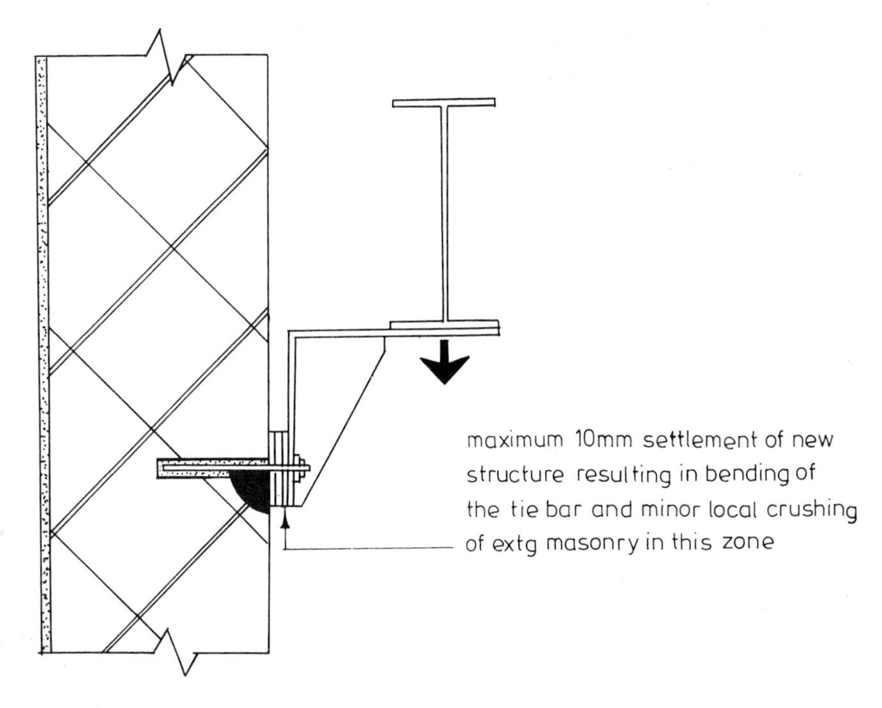

Figure 7.8 118–120 Colmore Row, Birmingham: effects of settlement of new structure.

Figure 7.9 118–120 Colmore Row, Birmingham: after redevelopment.

Figure 7.10 118–120 Colmore Row, Birmingham: after redevelopment.

Figure 7.11 118–120 Colmore Row, Birmingham: elements of temporary support system threaded into the existing structure.

Figure 7.12 118–120 Colmore Row, Birmingham: view from rear showing elements of temporary support system during demolition work.

Figure 7.13 118–120 Colmore Row, Birmingham: new beam and facade-tie brackets in permanent positions.

Figure 7.14 118–120 Colmore Row, Birmingham: aerial view showing retained facade and new steel frame.

Case study two

St Paul's House, Leeds

Background to the scheme

St Paul's House, a Grade II listed building, was designed by Thomas Ambler and built in 1878 as a warehouse for John Barran, a Leeds clothing manufacturer. The building, widely recognized as the best example of the nineteenth century use of an Hispano-Moorish architectural style in England, comprised brick and terra-cotta elevations surmounted by an ornate parapet perforated by roundels, and five terra-cotta minarets. Due to decay and instability, the whole of this ornate parapet, along with the corner minarets, had to be removed in the early 1960s. The external load-bearing walls were founded on strip footings of sandstone blocks, and the internal structure comprised cast iron columns on a 3.90 × 5.49 m grid supporting steel main and secondary beams with substantial timber floors. The building is located in Park Square, Leeds, probably the best Georgian square in the city, and therefore of considerable historic and environmental importance.

The principal reason for the Grade II listing of St Paul's House was its elaborate facade, none of its internal structure and fabric being of particular merit. In addition, the structure and layout of the existing interior were found to be unsuitable for adaptation to provide the modern lettable office accommodation required by the client, and the insertion of an entirely new structure within the retained listed facade proved to be the only viable solution. This compromise, which satisfied the client's needs, whilst retaining the historically and architecturally important elements of the building, raised no objections from local conservationists, and planning permission and listed building consent were duly granted.

The new *in situ* reinforced concrete structure erected within the retained facade comprises flush slab floors supported by square mushroom-headed columns on a grid of 7.8 × 5.5 m. This new structure provided two additional floors, one between the original ground and first floor levels, and the other at the original roof level, necessitating the addition of a new mansard roof. The scheme also included a major restoration of the facade which involved complete reinstatement of the original ornate parapet and minarets using glass reinforced plastic replicas. Figures 7.15 and 7.16 show a plan and cross-section of the scheme, and Figures 7.20 and 7.21 show the building after completion.

Temporary support system

Initially, an external support system was considered since this would have allowed the demolition and new construction to be carried out unhindered by internal temporary supports. However, due to the relatively narrow streets bordering three sides of St Paul's House, an external support system was ruled out, since it would have caused severe disruption to existing pedestrian and traffic flow. The final choice, designed by the main contractor and shown in Figure 7.17, comprised a wholly internal support system, consisting of four structural steel towers designed to support the facade walls via scaffold-tube flying shores at two levels. The towers comprised standard steel channel components, with diagonal and horizontal bracing, located centrally in line along the main axis of the building. The towers were founded on reinforced concrete pads, each incorporating four macalloy bar rock anchors to a depth of 18 m. The flying shores, constructed from standard scaffolding components, were 1.2 m square in cross-section, with diagonal bracings. These shores supported the facade by means of twin timber walings on each side of the walls which acted as a collar around the facade. The walls were rigidly held between this collar of walings using hardwood folding wedges (detail A, Figure 7.17). A total of eight sets of flying shores with their timber walings were inserted at two levels within the existing structure.

The towers and shores had to be constructed before any demolition took place and this was effected by pre-assembling the towers and lowering them into place by tower crane through holes previously formed in the existing roof and floors. Care had to be taken to ensure that no existing main structural members were cut or weakened during the installation of the towers. The two levels of flying shores were positioned so that they occurred between existing floor levels and could be assembled from the existing floors. In addition to ensuring that the support system did not conflict with the existing structure, it was also essential to ensure that it did not conflict with the new, and this further affected the location of every component of the towers and shoring. Important factors included ensuring that the diagonal bracing to the towers did not intersect the new floor slabs and that the two levels of flying shoring did not interfere with the construction of the new columns and floors. Figure 7.17 shows the relationship between the temporary support system and the existing and new structures, and Figure 7.22 shows one of the support towers erected within the existing building prior to its demolition.

The erection sequence for the temporary support system was as follows.

1. drill ground anchors for tower bases;
2. cast reinforced concrete pad foundations for towers, including holding down bolts;
3. stress ground anchors;
4. remove all existing windows and parts of floors for insertion of towers;
5. fix twin walings along each side of walls using bolts (detail A, Figure 7.17);
6. erect central support towers;
7. construct flying shores and connect to walings.

Following the erection of the support system and the securing of the facade to it, the existing building's interior was demolished and construction of the new reinforced concrete structure commenced. As each new floor was cast, four small pockets, 400 mm square, were left around the legs of the support towers and, on completion, the legs were drawn out by tower crane, the remaining holes being made good. Figures 7.23 and 7.24 show parts of the temporary support system after the demolition phase and during construction of the new structure.

Facade ties

The St Paul's House facade was tied back to the new structure using steel angles bolted to the underside of the new floor slab edges and resin-anchored into the facade masonry using pumped resinous mortar. The steel angles varied in length from 475–950 mm and were located at 2.7–3.9 m intervals along the floor slab edges. Their locations are shown in Figure 7.15. Because the existing walls comprised only thin outer skins of good masonry with very insubstantial, uncompacted rubble infill, the use of resin cartridges was ruled out due to the possibility of loss of the limited amount of resinous mortar by seepage into the cavitated masonry. Instead, pre-mixed resinous mortar was pumped into the pre-drilled holes, allowing seepage into the surrounding masonry where necessary, until the holes were seen to be full. The tie-bars were then pushed into these resin-filled holes, anchoring the steel angles to the facade. The steel angle facade-ties, which are detailed in Figure 7.18, were fixed in the following manner.

1. Two or three holes (depending on size of steel angle), 320 mm deep

× 25 mm diameter, drilled into facade masonry to take 'Lockset-Rebar' tie-bars.

2. CBP 'Lokset' type 'P' pre-mixed resinous mortar pumped into holes in masonry. Pre-mixed resinous mortar can be used as an alternative to resin cartridges (section 6.1.2). A polyester resin composition is mixed on site with a catalyst to form the resinous mortar which is pumped into the fixing hole prior to inserting the anchor. This method is often considered preferable where the masonry might contain internal cracks or cavities into which the limited amount of resinous mortar produced by a cartridge would be lost due to seepage. The pre-mixed resinous mortar can be pumped into the anchor hole until there is visible evidence that it is completely filled before the anchor is inserted.
3. Steel angle positioned with the 70 × 30 mm slots in its vertical leg over the resin-filled holes; 450 mm long × 20 mm diameter tie-bars pushed through slots into the resinous mortar with threaded ends left projecting; angle secured to facade with nuts.
4. Vertical bolts for fixing angle to underside of new floor slab inserted through horizontal leg and held in position by twin nuts. (An angle at this stage of the fixing process is shown in Figure 7.24 – bottom left corner).
5. Floor slab soffit formwork erected and reinforcement fixed.
6. Floor slab concrete poured and compacted around connection bolts to complete the fixing.

Differential settlement

The following precautions, detailed in Figure 7.18, were taken to ensure that no physical damage was caused to the facade-ties or to the new and existing structures in the event of settlement of the new building.

1. The facade-tie angles were secured to the masonry using 20 mm diameter tie-bars passed through the bottom of 70 mm deep slots in the angles' vertical legs, and the fixing nuts were hand-tightened. This, together with the provision of four layers of slip-strip between the mild steel fixing plates and the angles, ensured that the new structure could settle by up to 50 mm without placing stress on the facade-ties, the floor slab edges, or the existing masonry. (This technique is described fully in section 6.1.3.)
2. The insertion of a 10 mm strip of compressible filler at the interface of the new floor slab edges and the facade ensured that no bond was formed and permitted settlement of the new structure to take place unrestricted and without the risk of damage.

Foundation design

In order to minimize disturbance to the existing strip footings, the outside lines of new columns were positioned some 2.5 m back from the retained facade, and the outside strip of each new floor slab cantilevered from these columns to its junction with the facade's inner face (Figure 7.16). This resulted in a separation between the edges of the existing footings and the new foundations sufficient to ensure that their excavation and construction did not adversely affect the existing sandstone strip footings and, therefore, the stability of the facade.

Reinstatement of original parapet and corner minarets using glass reinforced plastic

The original building was surmounted by an ornate brick and terra-cotta parapet and five terra-cotta minarets which gave the building an air of flamboyancy typical of its architectural style. In the early 1960s these features were removed due to structural instability. Fortunately one of the original minarets and a section of parapet had been kept in storage and it was decided that these should be used as patterns to make copies, so that the building could be restored to its original exterior design. In the first instance, quotations for copies in terra-cotta, the original material, were obtained but the cost was considered prohibitive. Ultimately a craftsman with the ability to imitate a range of natural materials, including terra-cotta, in glass reinforced plastic was commissioned to produce a full replica of the original parapet and minarets. The glass reinforced plastic contained special colouring pigments to match the existing terra-cotta and on the completed building the new parapet and minarets appear highly authentic to the passer-by. A few other badly eroded portions of terra-cotta were also replaced in glass reinforced plastic and the total cost of the work was considerably less than the estimated cost of using natural terra-cotta, thereby allowing this important piece of restoration work to be carried out.

A further problem was created by the light weight of the glass reinforced plastic minarets relative to their size (5.5 m high × maximum 1.8 m wide). A stable anchorage had to be provided capable of resisting the high wind pressures which would inevitably occur at roof level. The solution was to fix a 200 mm square hollow section column in each minaret position. The hollow glass reinforced plastic mouldings were placed over the columns and secured to them via eight steel angle connections at five levels. The connections were in some cases cemented, and in other cases bolted to the glass reinforced plastic.

The problem of anchoring the minaret at the north-east corner of the building led to a quite unique solution to another design problem. The local authority would not permit the boiler flue to extend to its required height above roof level because of the adverse effect on the appearance of the building when viewed from a distance. This presented a serious problem, which by virtue of the hollow construction and height of the minarets, was solved very simply. The boiler flue was taken up to the base of the north-east minaret, and instead of using a 200 mm square hollow section column to provide support, a 450 mm diameter hollow section, which acted as the flue extension, was used. The minaret was anchored to the flue in the same way as the others via steel angles welded to the flue at their inner ends, and cemented or bolted to the glass reinforced plastic at their outer ends. Clearly, the domed top to the minaret had to be left open for emission of the flue gases. Details of the glass reinforced plastic restoration work are shown in Figure 7.19.

Architects Booth Shaw and Partners, Leeds.

Consulting engineers Alistair McCowan & Partners, Pontefract, West Yorkshire.

Main contractor Fairclough Building Ltd, Leeds.

GRP craftsman H. Butcher, Creeting St Mary, Suffolk.

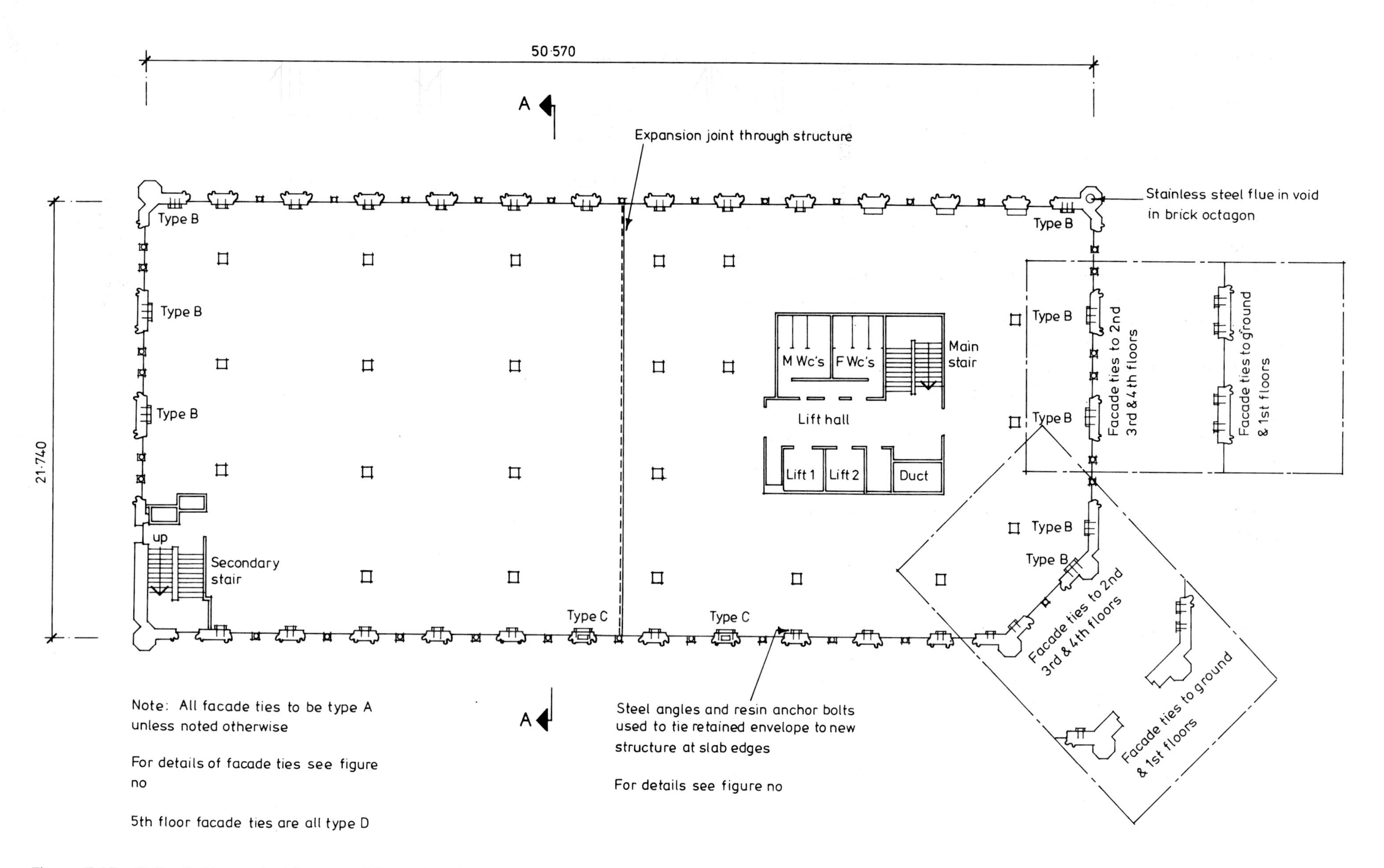

Figure 7.15 St Paul's House, Leeds: second floor plan showing locations of facade-ties.

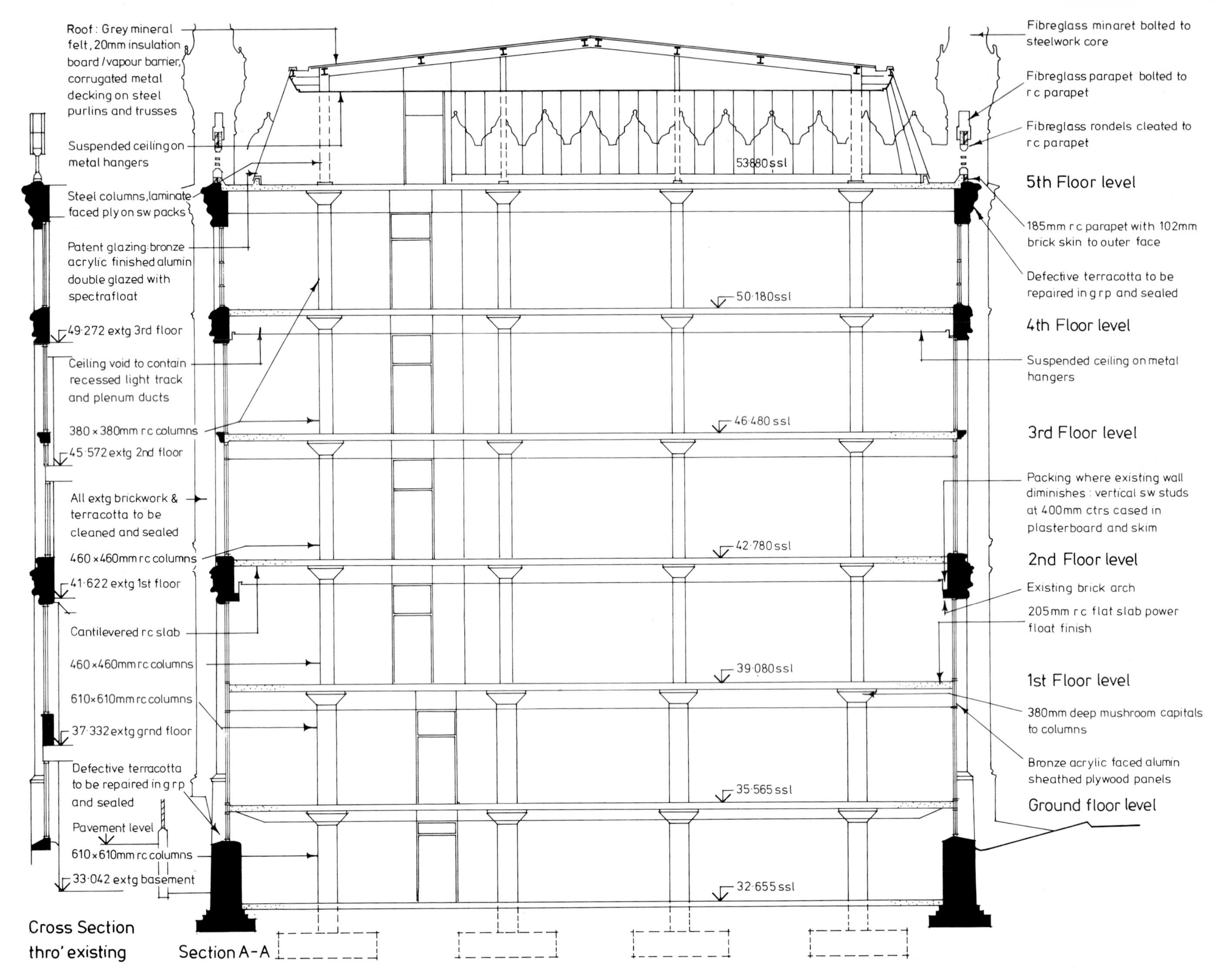

Figure 7.16 St Paul's House, Leeds: vertical cross-section.

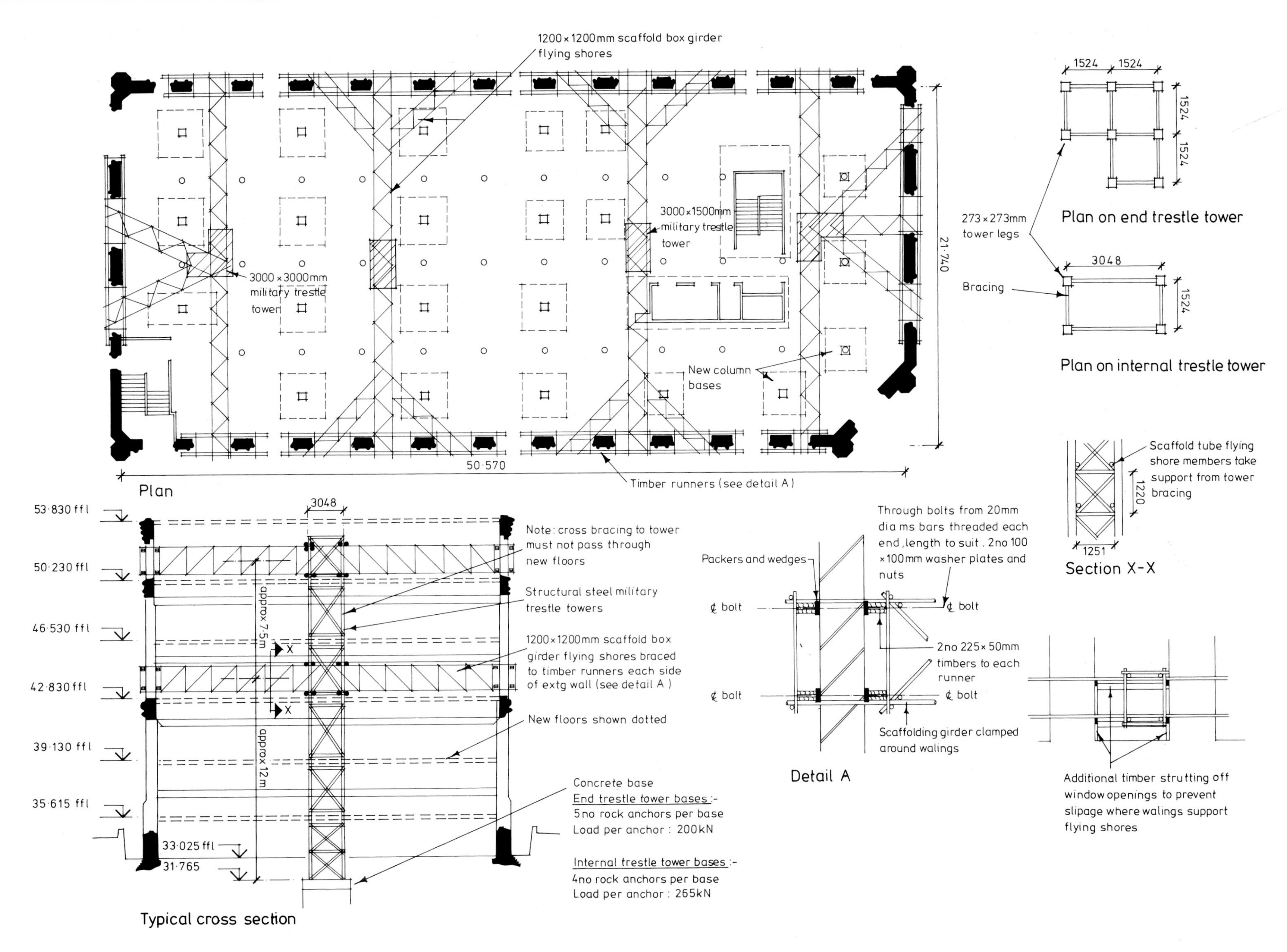

Figure 7.17 St Paul's House, Leeds: temporary support system.

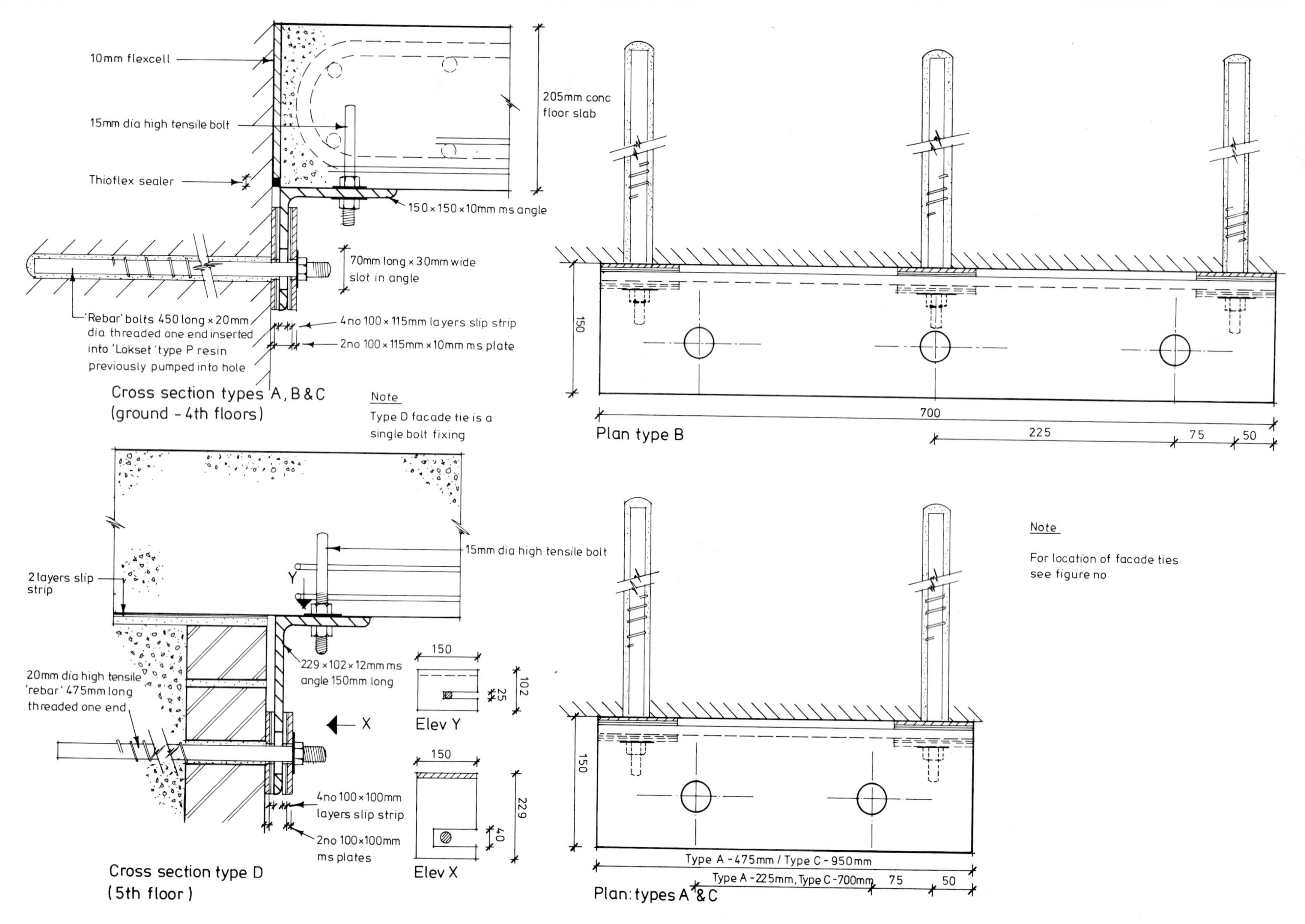

Figure 7.18 St Paul's House, Leeds: details of facade-ties.

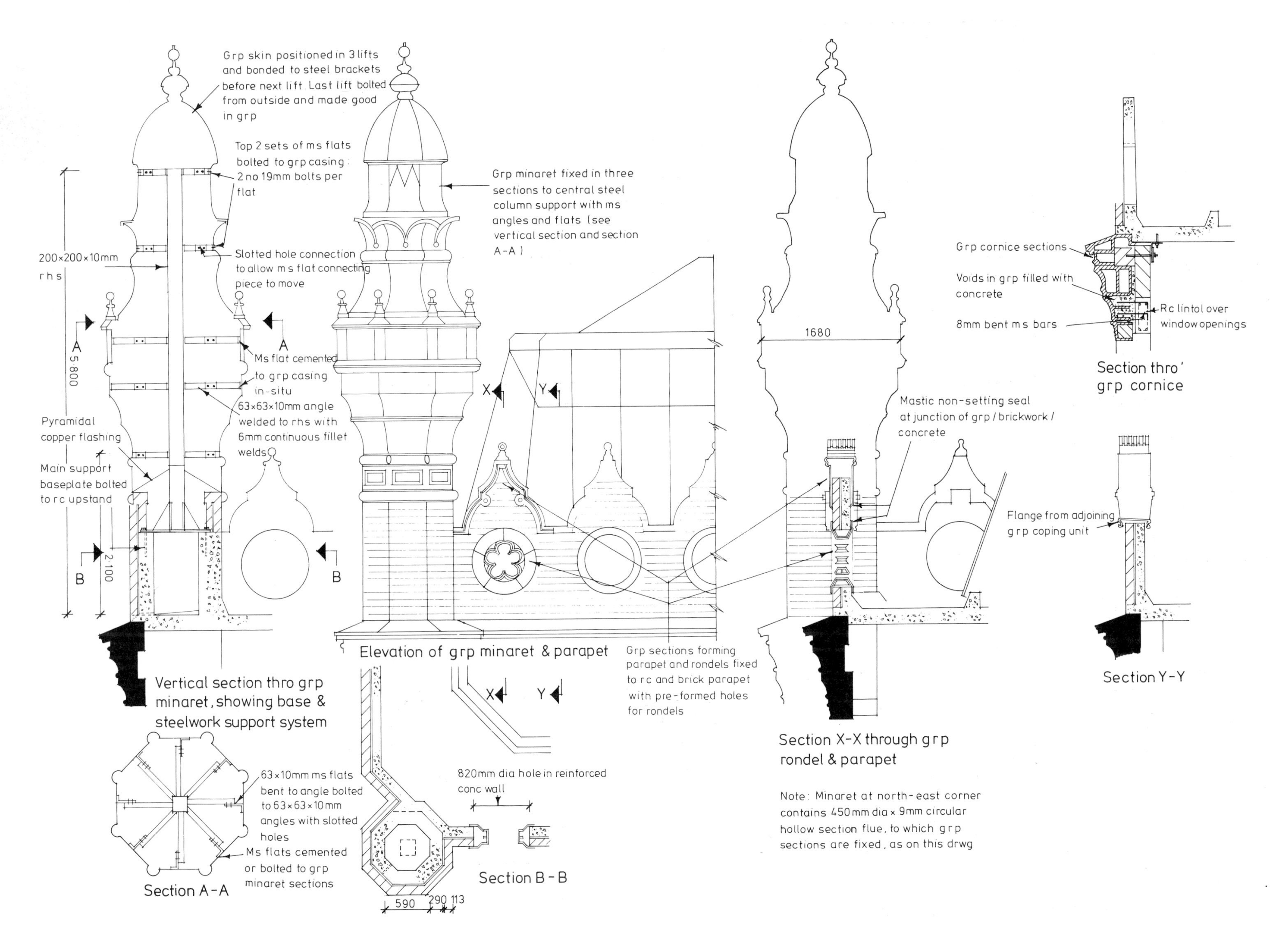

Figure 7.19 St Paul's House, Leeds: details of grp minarets, parapet and roundels.

Figure 7.20 St Paul's House, Leeds: after redevelopment.

Figure 7.21 St Paul's House, Leeds: after redevelopment.

Figure 7.22 St Paul's House, Leeds: elements of temporary support system erected within the existing structure.

Figure 7.23 St Paul's House, Leeds: interior view showing temporary support towers and flying shores.

Figure 7.24 St Paul's House, Leeds: interior view showing elements of temporary support system and construction of the new structure. (Note facade-tie angle in bottom left-hand corner.)

Figure 7.25 St Paul's House, Leeds: interior view after structural completion.

Case study three

Lloyds Bank, Harrogate

Background to the scheme

The buildings affected by the development, numbers 10 and 11 Cambridge Crescent, form part of a unified group of Grade II listed buildings from 1–12 Cambridge Crescent, designed by George Dawson in 1880. The Crescent, composed of twelve three-bay frontages stepped downhill from number 12 to number 1, comprises three storeys with cellars and attics roofed by slated mansards. Figure 7.34 shows numbers 9, 10, 11 and part of number 12 before the scheme commenced, and Figure 7.26 shows the existing front elevation. The gritstone Ashlar facade of numbers 10 and 11 incorporates red brick dressings and coarse florid detailing, and is surmounted by a bracketed cornice with round headed attic windows in the parapet above. The attic and second floor windows are arched sashes, the latter comprising two lights in recesses with cast iron balconies. The first floor windows, also arched, have enriched surrounds and cast iron balconies, the frames having been replaced by full height opening casements. The existing ground floor frontage, like most of the others in the Crescent, consisted of large glazed shopfronts.

The development of numbers 10 and 11 Cambridge Crescent involved demolition of the whole of the existing structure with the exception of the curved facade which was retained to conceal a new framed structure erected behind. The existing parapet and attic windows were not retained *in situ* like the remainder of the facade, but were dismantled stone by stone and stored for re-erection after completion of the new structure. An important consideration in the design was to maintain the original roof line behind the re-erected dormers. This was achieved by providing pitched roof beams as part of the new structural frame and a covering of slates to match the profile and character of the existing roofs in the Crescent. The most significant alteration to the retained facade was the replacement of the existing shopfronts with a modern six-bay bank frontage as shown in Figures 7.27 and 7.35. The six bays of this new frontage were divided by structural reinforced concrete piers and spanned by six beams supporting the facade above. Work below ground level included lowering of the original basement level to form a deeper basement to house a strongroom and storage areas. The new structure comprises a steel frame supporting *in situ* reinforced concrete floors. Unlike most facade retention schemes, where additional floors are usually inserted, the design incorporated the same number of floor levels as originally existed. Typical floor plans and a cross-section of the scheme are shown in Figures 7.28 and 7.29.

Temporary support system

The part external/part internal temporary support system to the retained facade, shown in Figures 7.30–7.32 and 7.36–7.39, was constructed from tubular scaffolding and comprised the following elements.

1. Two pairs of curved horizontal compound beams used to 'collar' the facade.

2. Light frameworks on each side of the facade, used to support the curved beams.
3. Two vertical cantilevers located in the stairwell positions used to restrain any lateral movement of the curved beams and therefore the facade.
4. Horizontal ties between the vertical cantilevers and the curved beams, extended to supporting towers at the rear of the building.
5. Two pairs of horizontal flying shores used to prevent inward movement of the existing party walls. In addition, the front pair of shores also acted as secondary supports to the two vertical cantilevers.

The figures show that most of the temporary support system's elements were located inside the building, with only two of the curved horizontal beams and their supports located outside the facade. As with all support systems consisting predominantly of internal elements, its erection was complicated by having to ensure that none of its components conflicted with either the existing or the new structures. Figures 7.38 and 7.39 show the integration of some of the temporary supports with the new structure.

The principal elements providing direct support to the facade were the two pairs of curved horizontal beams, the remaining scaffolding structure acting as secondary supports to them. The facade was collared between each pair of these curved box-section beams at two levels by means of horizontal adjustable reveal pins thrusting against timber wall plates on both sides of the wall (Figures 7.30 and 7.41). The curved horizontal beams, constructed from specially made curved scaffold tubing, received vertical support from the light scaffolding frameworks on each side of the facade, and lateral support from the two vertical cantilevers located in the new stairwells. The main function of the vertical cantilevers, which were the most important of the support system's secondary elements, was to restrain any lateral movement of the facade caused by wind loads on both its faces.

Facade ties

Two types of facade tie, both of which employed resin-anchors, were used to secure the facade back to the new structure. The first type, used to tie the facade directly to the new floor slab edges, employed 800 mm long × 16 mm diameter tie bars resin-anchored into the masonry and cast into the edges of the new *in situ* concrete floor slabs. The installation of these ties, shown in Figures 7.33 and 7.40, is described below.

1. 25 mm diameter holes drilled into the facade masonry at 750 mm centres along lines of new floor slab edges.
2. 22 mm diameter Celtite 'Selfix' resin capsule inserted into each hole. The sealed capsule contains a polyester resin composition with a measured proportion of hardener which, when the capsule is broken and they are mixed, form a rapid-setting resinous mortar. The hole depth and capsule length used for each tie varied between 200 mm and 400 mm according to the thickness of masonry available.
3. 800 mm long × 16 mm diameter hot rolled, sheradized deformed tie bar spun into the resin capsule inside hole using power tool. This operation breaks the capsule and mixes its ingredients to form the rapid-setting resinous mortar.

Operations 1 to 3 were carried out from the floor slab formwork, which was erected prior to the installation of the ties to act as a working platform.

4. Resinous mortar left to harden, anchoring tie bar into the facade masonry and leaving the outer end of the bar projecting into the new slab edge zone.
5. Reinforcement fixed and *in situ* concrete to new floor slabs poured and compacted around the projecting resin-anchored tie bars.

Similar resin-anchored ties were also used between the new floor slab edges and the existing party walls along both sides of the building.

The second type of facade tie was used to secure the facade to the new structural steel frame. The facade was indirectly tied to seven stanchions located immediately behind its inner face, by means of steel angle cleats, resin-anchored to the masonry and welded to the stanchions' flanges. These ties, shown in Figures 7.33 and 7.41, were installed in the following manner:

1. 25 mm diameter holes drilled into the facade masonry on both sides of perimeter stanchions at one metre vertical centres;
2. 22 mm diameter Celtite 'Selfix' resin capsule inserted into each hole;
3. 16 mm diameter hot rolled, sheradized deformed tie bar, threaded at outer end to receive fixing nut, spun into the resin capsule inside each hole using power tool;
4. steel angle cleat with pre-drilled hole in one leg, located over projecting threaded end of resin-anchored tie bar, leaving other leg in contact with stanchion flange;
5. leg of steel angle cleat welded to stanchion flange to hold the anchored leg hard against the wall face;
6. connection completed by locating and tightening nut over projecting end of resin-anchored tie bar.

Differential settlement and foundation design

No provisions were made, in either the facade ties or the interface detail, for the occurrence of differential settlement between the retained facade and the new structure. Both types of facade tie produced fully rigid connections, and no slip-surface was incorporated at the interface between the two structures. The lowering of the existing basement level required the retained facade to be underpinned with new mass concrete foundations, which, it was predicted, would cause the facade to undergo some settlement. Both the underpinning and the new structure's foundations were designed so that their settlement would be compatible, therefore resulting in equal, rather than differential, settlement of the facade and new structure. This, in turn, allowed the design of rigid facade ties since the equal settlement of the two structures ensured that they would not be subjected to stresses likely to result in damage to the ties themselves, the facade or the new structure. The equal settlement of the two structures also obviated the necessity of providing a slip-surface at their interface.

Architects Lloyds Bank, Architects Department, Leeds.

Consulting engineers F.R. Jenks & Partners, Sheffield.

Main contractor Walter Thompson (Contractors) Ltd, Northallerton, North Yorkshire.

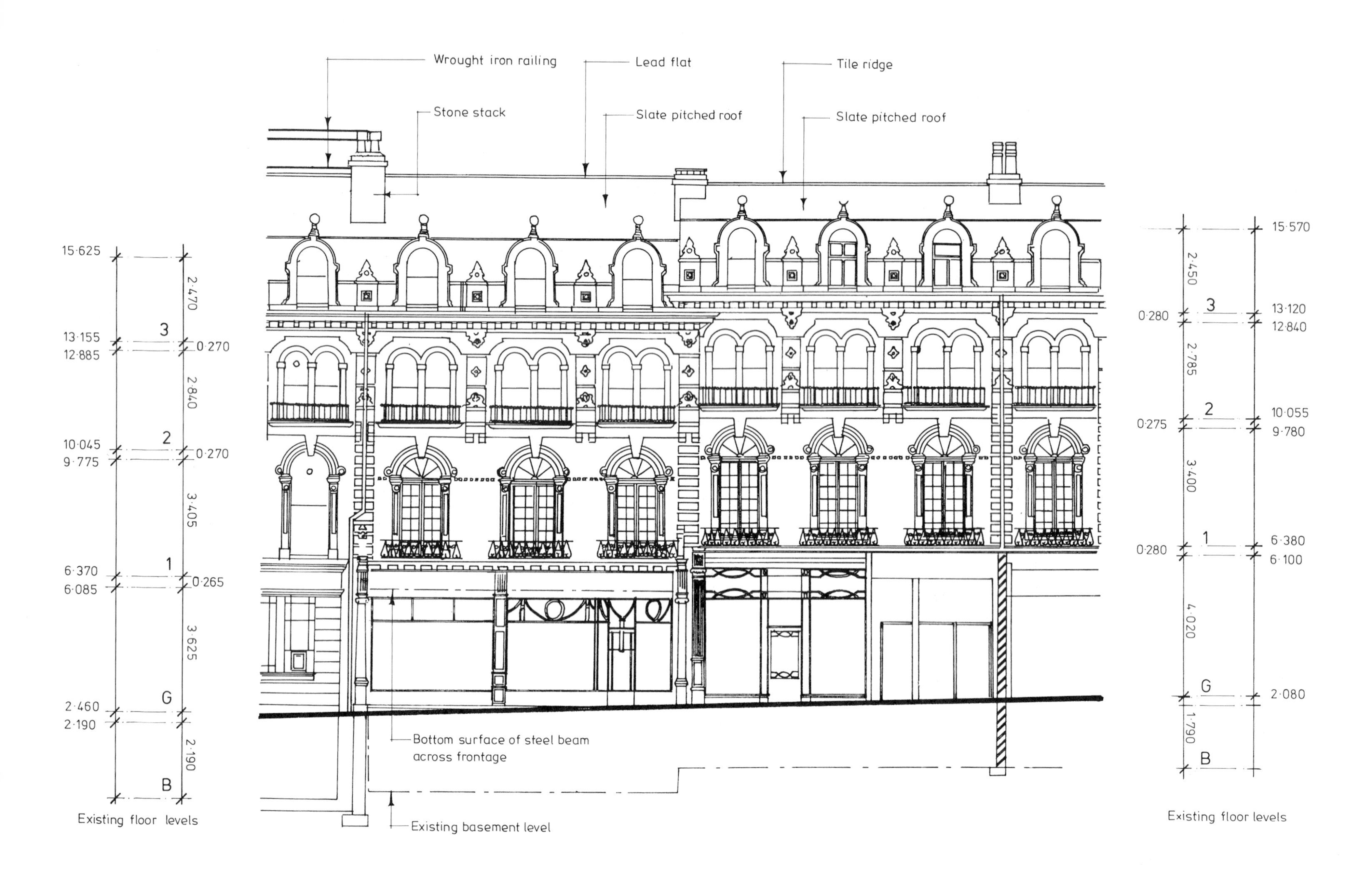

Figure 7.26 Lloyds Bank, Harrogate: front elevation before redevelopment.

Figure 7.27 Lloyds Bank, Harrogate: front elevation after redevelopment.

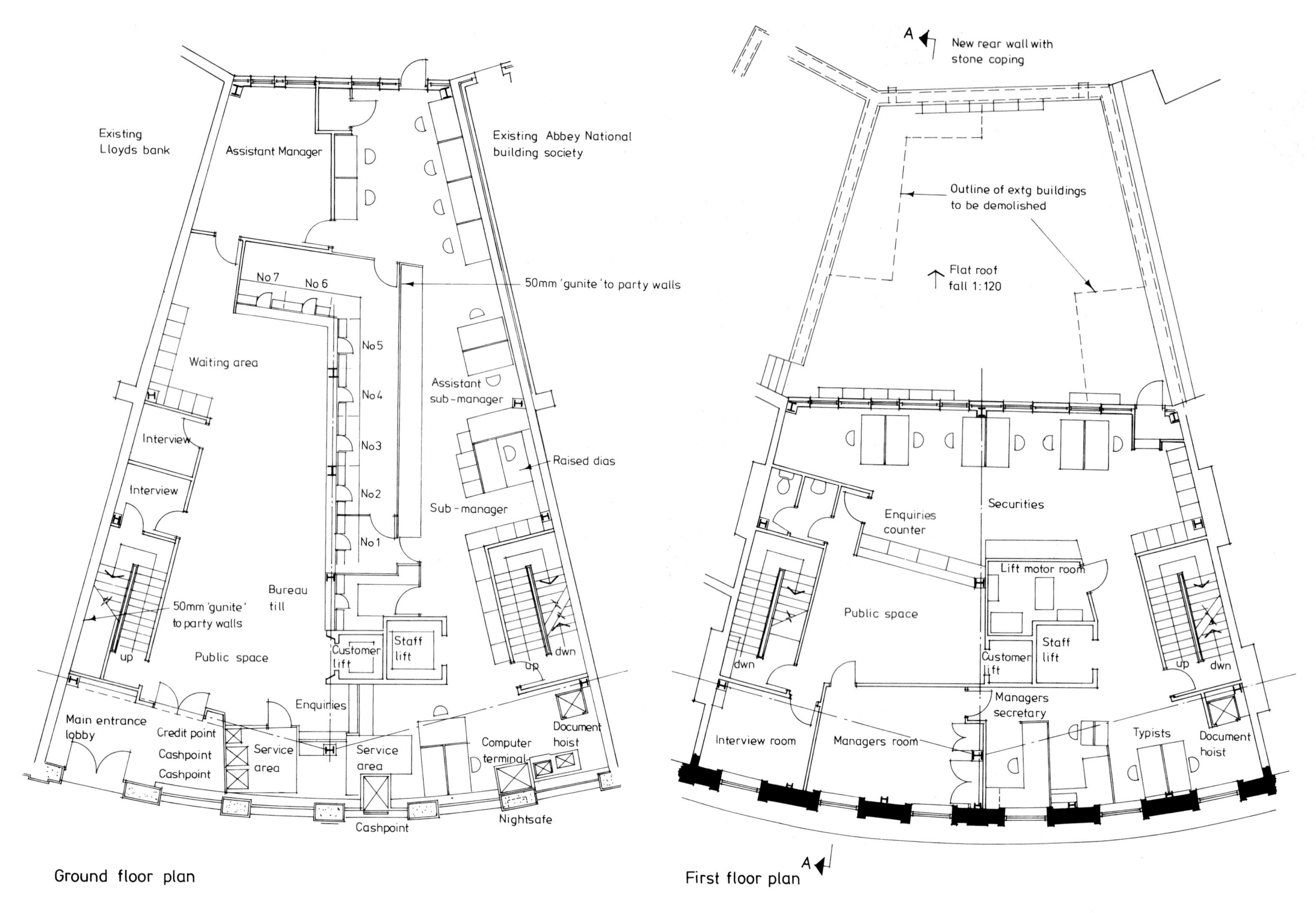

Figure 7.28 Lloyds Bank, Harrogate: ground and first floor plans.

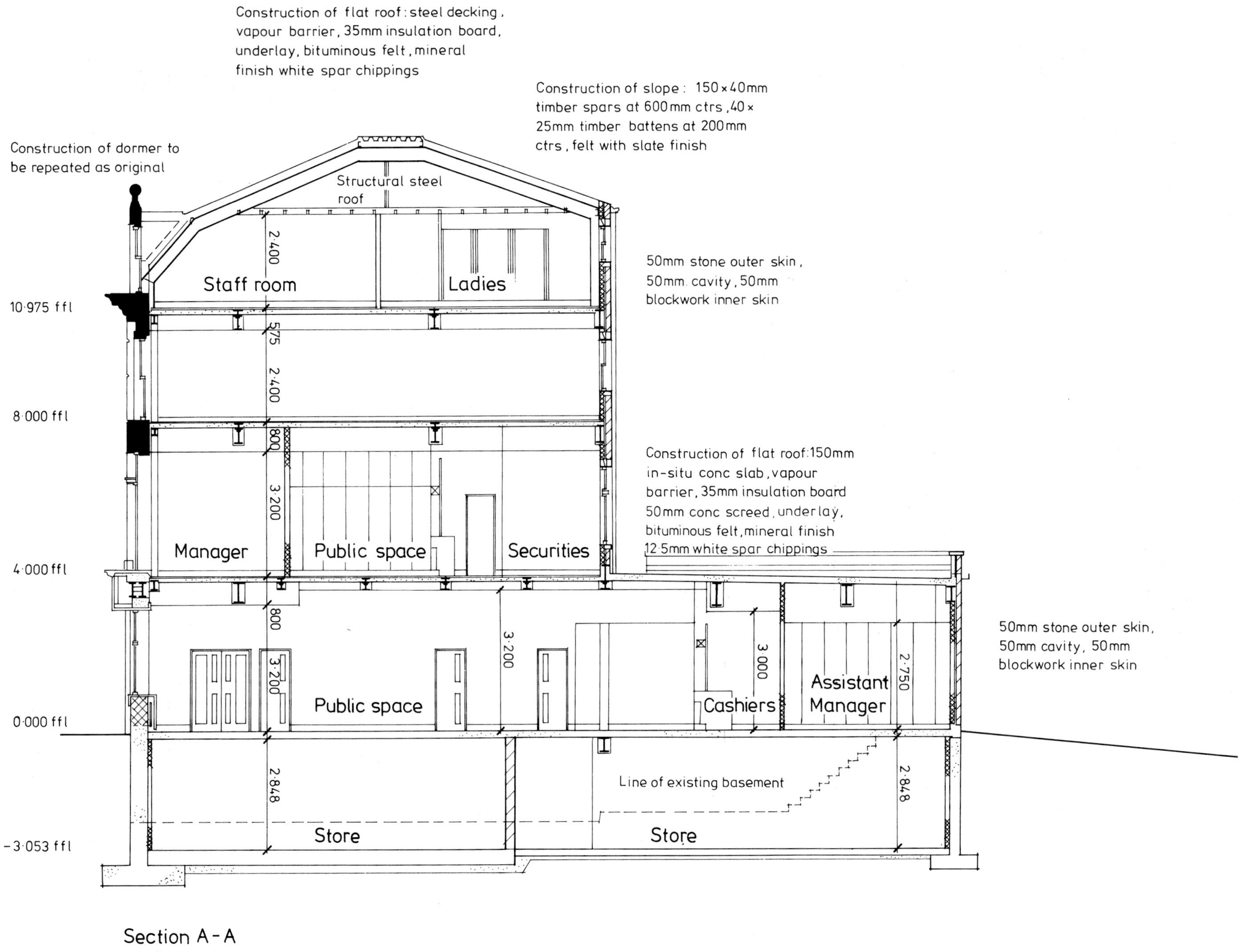

Construction of flat roof: steel decking, vapour barrier, 35mm insulation board, underlay, bituminous felt, mineral finish white spar chippings

Construction of slope: 150 × 40mm timber spars at 600mm ctrs, 40 × 25mm timber battens at 200mm ctrs, felt with slate finish

Construction of dormer to be repeated as original

Structural steel roof

Staff room

Ladies

2·400

50mm stone outer skin, 50mm cavity, 50mm blockwork inner skin

10·975 ffl

575

2·400

8·000 ffl

800

3·200

Construction of flat roof: 150mm in-situ conc slab, vapour barrier, 35mm insulation board 50mm conc screed, underlay, bituminous felt, mineral finish 12·5mm white spar chippings

Manager

Public space

Securities

4·000 ffl

800

3·200

3·200

3·000

2·750

50mm stone outer skin, 50mm cavity, 50mm blockwork inner skin

Public space

Cashiers

Assistant Manager

0·000 ffl

2·848

Line of existing basement

2·848

Store

Store

−3·053 ffl

Section A - A

Figure 7.29 Lloyds Bank, Harrogate: vertical cross-section.

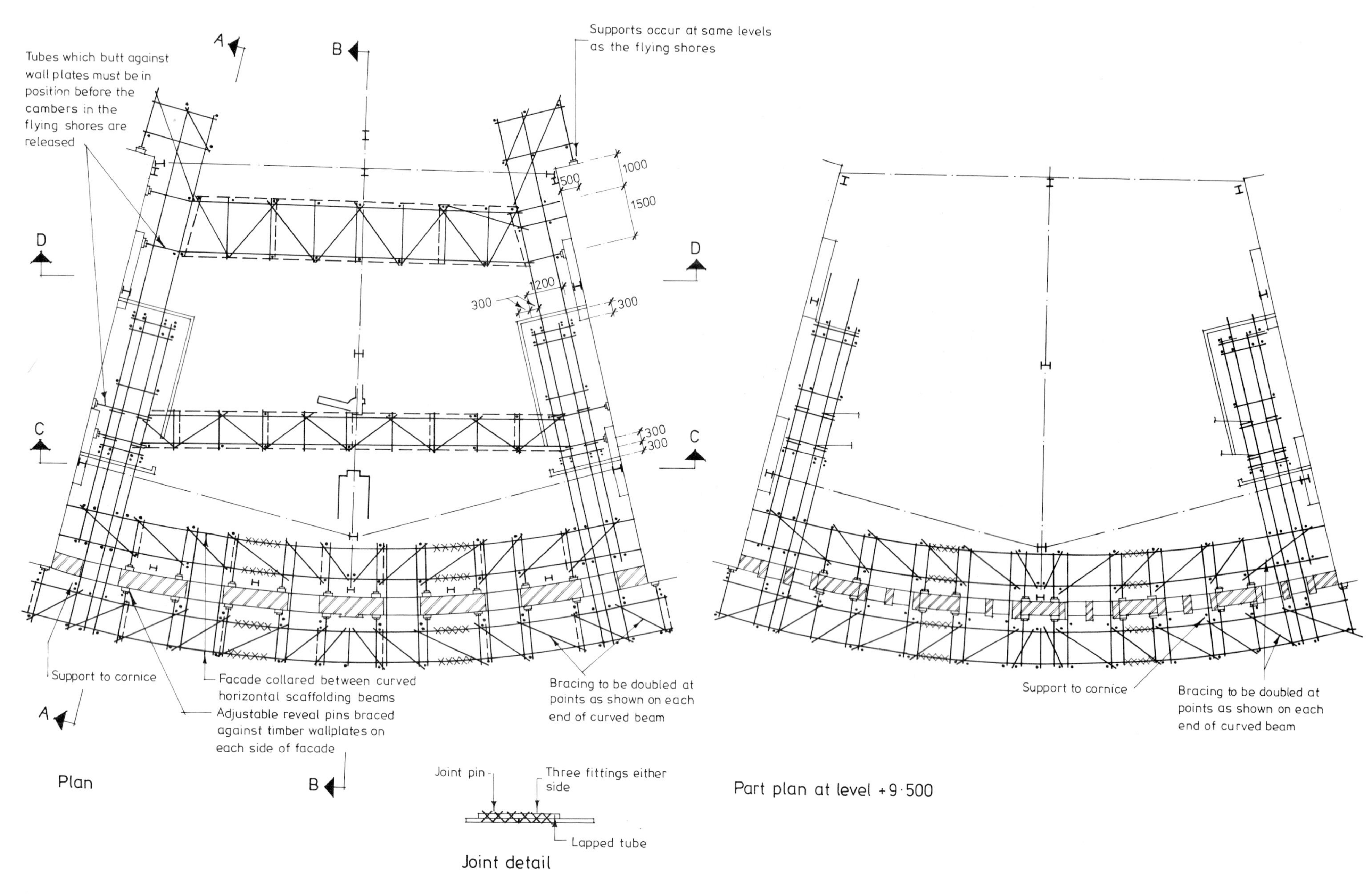

Figure 7.30 Lloyds Bank, Harrogate: temporary support system.

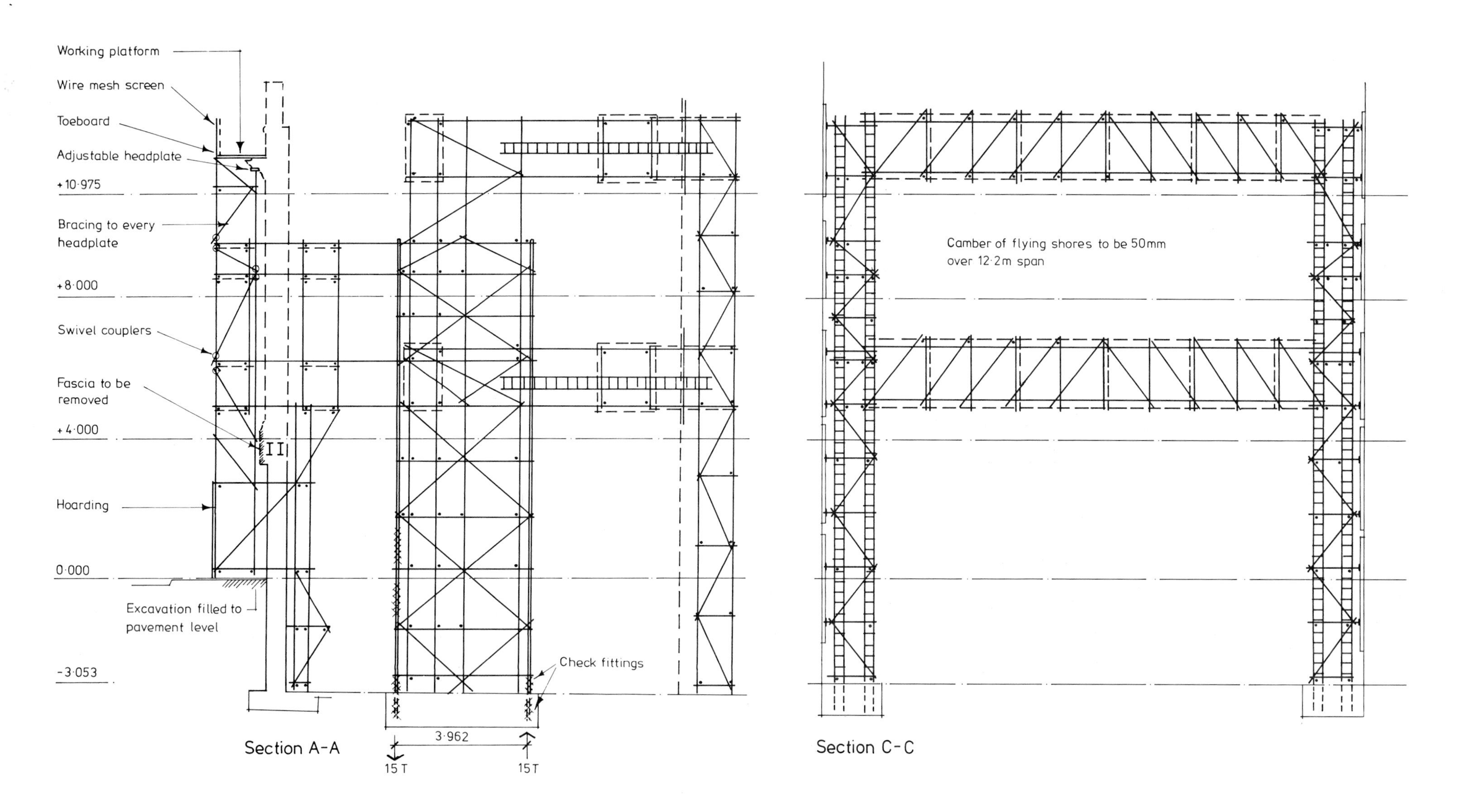

Figure 7.31 Lloyds Bank, Harrogate: temporary support system.

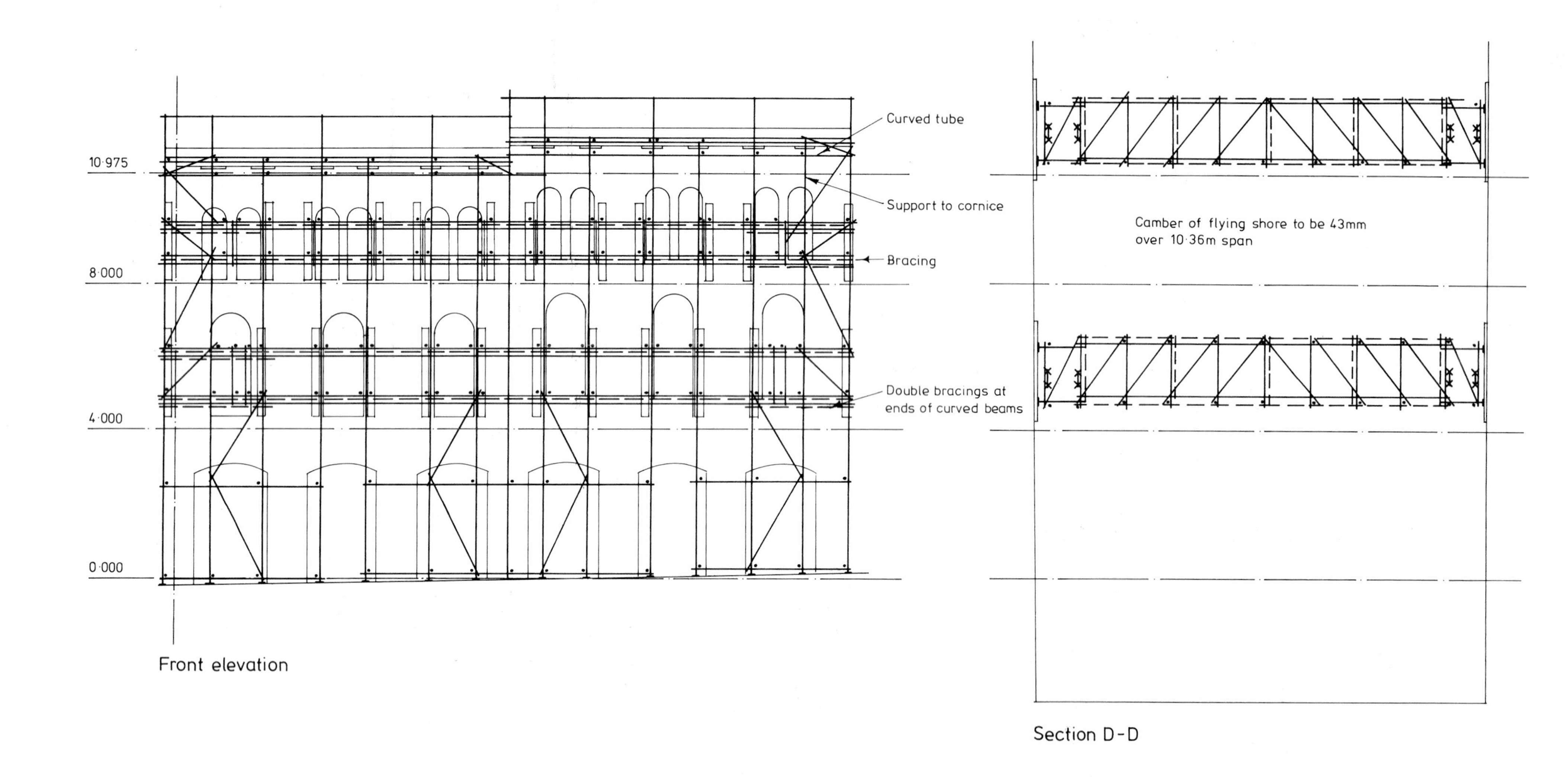

Figure 7.32 Lloyds Bank, Harrogate: temporary support system.

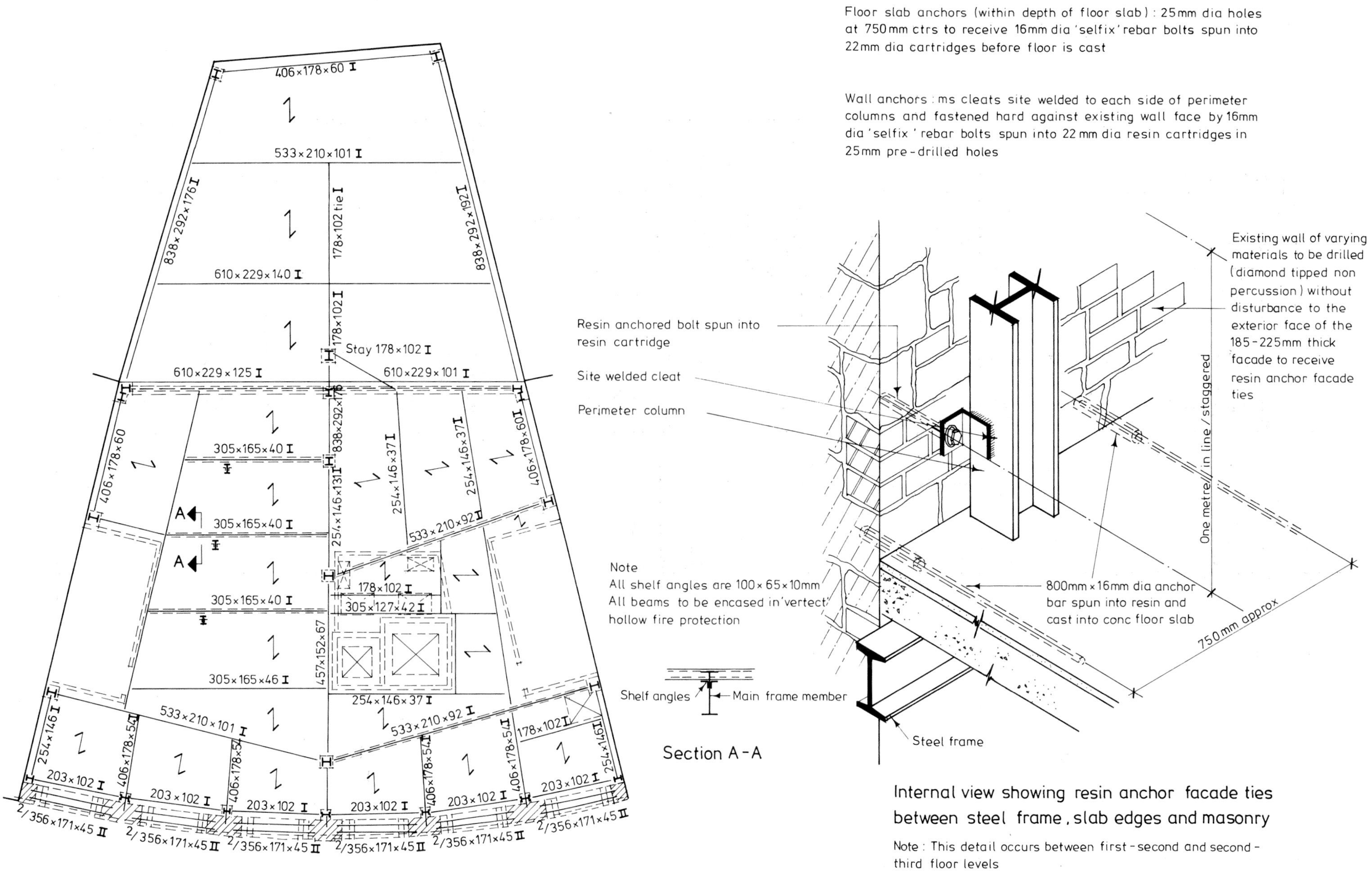

Floor slab anchors (within depth of floor slab): 25mm dia holes at 750mm ctrs to receive 16mm dia 'selfix' rebar bolts spun into 22mm dia cartridges before floor is cast
Wall anchors: ms cleats site welded to each side of perimeter columns and fastened hard against existing wall face by 16mm dia 'selfix' rebar bolts spun into 22mm dia resin cartridges in 25mm pre-drilled holes
406×178×60 I
533×210×101 I
838×292×176 I
178×102 tie I
838×292×192 I
610×229×140 I
178×102 I
Stay 178×102 I
610×229×125 I
610×229×101 I
406×178×60
838×292×176
305×165×40 I
254×146×37 I
254×146×37 I
406×178×60 I
254×146×131 I
305×165×40 I
533×210×92 I
305×165×40 I
178×102 I
305×127×42 I
457×152×67
305×165×46 I
254×146×37 I
533×210×101 I
533×210×92 I
254×146 I
406×178×54
406×178×54
406×178×54
406×178×54
178×102 I
254×146 I
203×102 I
203×102 I
203×102 I
203×102 I
203×102 I
203×102 I
2/356×171×45 II
2/356×171×45 II
2/356×171×45 II
2/356×171×45 II
2/356×171×45 II
2/356×171×45 II
A
A
Resin anchored bolt spun into resin cartridge
Site welded cleat
Perimeter column
Existing wall of varying materials to be drilled (diamond tipped non percussion) without disturbance to the exterior face of the 185-225mm thick facade to receive resin anchor facade ties
One metre in line / staggered
Note
All shelf angles are 100×65×10mm
All beams to be encased in 'vertect' hollow fire protection
800mm×16mm dia anchor bar spun into resin and cast into conc floor slab
750mm approx
Shelf angles
Main frame member
Section A-A
Steel frame

Internal view showing resin anchor facade ties between steel frame, slab edges and masonry

Note: This detail occurs between first-second and second-third floor levels

Plan at first floor level +4·0m showing structural steelwork

Figure 7.33 Lloyds Bank, Harrogate: new steel frame and facade-ties.

Figure 7.34 Lloyds Bank, Harrogate: before redevelopment.

Figure 7.35 Lloyds Bank, Harrogate: after redevelopment.

Figure 7.36 Lloyds Bank, Harrogate: external elements of temporary support system.

Figure 7.37 Lloyds Bank, Harrogate: view from rear after demolition, showing internal elements of temporary support system.

Figure 7.38 Lloyds Bank, Harrogate: view from rear showing temporary support system and new steel frame.

Figure 7.39 Lloyds Bank, Harrogate: view from rear showing uppermost curved compound scaffold beam supporting the facade.

Figure 7.40 Lloyds Bank, Harrogate: resin-anchored tie bars between retained facade and new floor slab, prior to pouring concrete.

Figure 7.41 Lloyds Bank, Harrogate: interior view showing horizontal reveal pins collaring the facade, and resin-anchored angle cleats securing the facade to the front stanchions.

Case study four

144 West George Street, Glasgow

Background to the scheme

The Category B listed building was designed by local architect James Sellars and built between 1879 and 1881 as premises for the New Club, a business and professional club. Most of the detailed architectural interest of the facade is concentrated on the ground and first floors, whose united composition is emphasised by the two storey portico and the bow fronted western bay. The elegant mouldings and other low-level carvings on the facade are among the finest examples of Victorian Glasgow's craftsmanship in masonry sculpture. The three ground floor bull's eye windows are echoed above by five smaller windows in the mansard. The frontage of the second, third and fourth floors contain less sculpture than the lower facade and are more conventionally Glaswegian-classical. Soon after the Second World War, the New Club joined with the Western Club and the building was taken over by the Cornhill Insurance Company for use as offices, which resulted in a number of internal alterations, including the sub-division of several of the larger public rooms. The interior of the existing building was of no particular merit, the principal reason for its listing being its facade which comprises one of Glasgow's best examples of Victorian architecture. The facade is also an important feature in the existing townscape, thereby adding to its architectural and scenic value.

The re-development involved total demolition of the existing building's interior and the insertion of a new framed structure behind the preserved facade. The new structure, which comprised an *in situ* reinforced concrete frame and tee-beam floors, incorporated two additional floor levels, one occurring between the original second and third floor levels, and the other at the original roof level. The provision of the latter floor resulted in the addition of a new slate mansard directly above, and set back from the existing mansard. This represented the most significant alteration to the appearance of the retained facade and is shown in Figure 7.42. The new accommodation has natural ventilation, with opening windows, and a low pressure hot water heating system using cill height perimeter convector heaters. This ruled out the need to provide deep suspended ceilings to house bulky air-conditioning equipment, thereby simplifying the problem of new floor levels and ceilings conflicting with the existing facade's fenestration. Existing and proposed front elevations are shown in Figure 7.42, and Figure 7.49 shows the front elevation as the scheme neared completion.

Temporary support system

The temporary support system was similar in principle to that used at 118–120 Colmore Row, Birmingham (Case study one): part of the new building's structural frame being used, initially, to provide temporary support to the retained facade. A single bay structural steel framework, comprising four stanchions (which would ultimately form part of the new permanent structure) and four horizontal wind girders, was erected immediately behind the facade before any demolition of the existing interior took place. The facade was tied back to this temporary framework at four levels using temporary resin-anchor ties between the front 152 × 152 mm member of each wind girder and the facade masonry. A number of similar temporary resin-anchor ties were also used at the topmost level, where a 305 × 165 mm member was inserted to facilitate the connections. Figures 7.46 and 7.51–7.53 show the layout of the temporary support system, and Figure 7.47 shows various details, including a temporary resin-anchor tie. Figure 7.54 also shows some of the temporary resin-anchor ties which were made using CBP 'Lokset' resin cartridges and 'Lokset-Rebar' tie bars.

One of the problems in forming the temporary resin-anchor ties between the facade and its support system was the variation in the gap width between the 152 × 152 mm wind girder members and the facade masonry caused by the latter's unevenness. This was overcome by using steel packing plates to fill the gap and achieve a solid connection at each tie position (Figures 7.47 and 7.54). At other points, where the gap width was excessive and thin packing plates were unsuitable, short structural 'I' sections were used to achieve the connections (Figures 7.47, 7.53 and 7.54). When all of the temporary resin-anchor ties had been completed, and the facade secured to its temporary support system, demolition of the existing structure could commence.

As previously stated, the temporary support system ultimately formed part of the new structure, its four stanchions being encased in concrete and incorporated as elements of the new *in situ* reinforced concrete frame. The temporary support system's four horizontal wind girders were also retained as permanent structural elements by casting them into the new *in situ* concrete floor slabs (Figure 7.55). It was the front 152 × 152 mm members of these wind girders which, in addition to being used initially for the temporary resin-anchor ties, were used to secure most of the permanent resin-anchor facade ties back to the new structure. The only elements of the temporary support system not incorporated into the new structure were the vertical bracing members and the temporary resin-anchor ties, these being cut off using oxy-acetylene burning equipment. It should be made clear, however, that none of the temporary ties were disconnected until all of the permanent facade-ties had been installed.

Facade ties

The facade was permanently tied back to the new structure by means of resin-anchor ties between the new *in situ* concrete floor slab edges and the facade masonry. Two designs were employed for these permanent resin-anchor ties and these are detailed in Figure 7.48. Those at the wind girder positions (levels G, 2, 4 and 5), and at level 6, employed 25 mm wide × 6 mm thick galvanized steel flats resin-anchored into the masonry directly above the outer 152 × 152 mm horizontal wind bracing member and bent over it. The floor slab formwork acted as a working platform at each level for the fixing of these ties, after which the new slab was cast, encasing the complete wind bracing member and the facade ties. Figure 7.55 shows a wind bracing member and the floor formwork, with part of the reinforcement fixed, prior to the casting of the floor slab. The permanent resin-anchor ties at levels 1 and 3 also employed 25 mm wide × 6 mm thick galvanized steel flats resin-anchored into the masonry and cast into the new floor slab edges. Unlike the ties at the wind girder positions, these were straight and the lateral restraint they provided was increased by the provision of a 200 mm long × 10 mm diameter galvanized dowel inserted through the inner end of each flat. Ties of this type were also used at the other levels at positions where the gap between the outer 152 × 152 mm wind bracing member and the facade was excessive. The material used to anchor the steel flats into the facade masonry was CBP 'Lockset' type 'P' pre-mixed resinous mortar, this being pumped into pre-drilled holes in the masonry and the ties pushed into it. This type of resinous mortar was also used at St Paul's House, Leeds (Case study two, p. 59) and at Abbey House, Glasgow (Case study seven, p. 127).

Differential settlement

In order to rule out the possibility of damage being caused to the facade ties, the facade itself, or the new structure in the event of the latter's settlement, the following features, which are shown in Figure 7.48 were incorporated into their design.

1. The provision of a 75 mm cube or 20 mm strip of polystyrene to

each facade-tie (see details) ensured that any downward movement of the new structure would result only in compression of the polystyrene, rather than bending of the facade-tie and associated damage to the masonry.

2. The insertion of a 6 mm strip of compressible filler at the interface of the new concrete floor slab edges and the facade masonry ensured that a physical bond was not formed between the two, thereby allowing the new structure to move downwards in relation to the facade without causing damage to the masonry or the new structure. The same type of compressible filler was also inserted at the interface of the new structure's front columns and the retained facade.

Foundation design

As with the Colmore Row scheme in Birmingham (Case study one, p. 44), the use of part of the new structure as the temporary support system required columns to be located immediately adjacent to the retained facade. It was also considered undesirable to place the facade's stability at risk by undermining it in any way and this therefore resulted in the bases to the six front columns, which had to be constructed wholly inside the line of the facade, being eccentrically loaded (Figure 7.45). The problem was overcome using the same principle as that used at Colmore Row, Birmingham (Case study one, p. 44), and described and illustrated more fully in Chapter 6, by using balanced-base foundations. By structurally linking the eccentrically loaded front bases to axially loaded bases on the next grid-line, the overturning moments were counterbalanced and the risk of failure ruled out. A plan of these balanced-base foundations is shown in Figure 7.45.

Architects Scott, Brownrigg & Turner, Glasgow.

Consulting engineers Thorburn & Partners, Glasgow.

Main contractor Henry Boot Construction Ltd, Glasgow.

Proposed front elevation (additions shaded)

Existing front elevation

Figure 7.42 144 West George Street, Glasgow: front elevation before and after redevelopment.

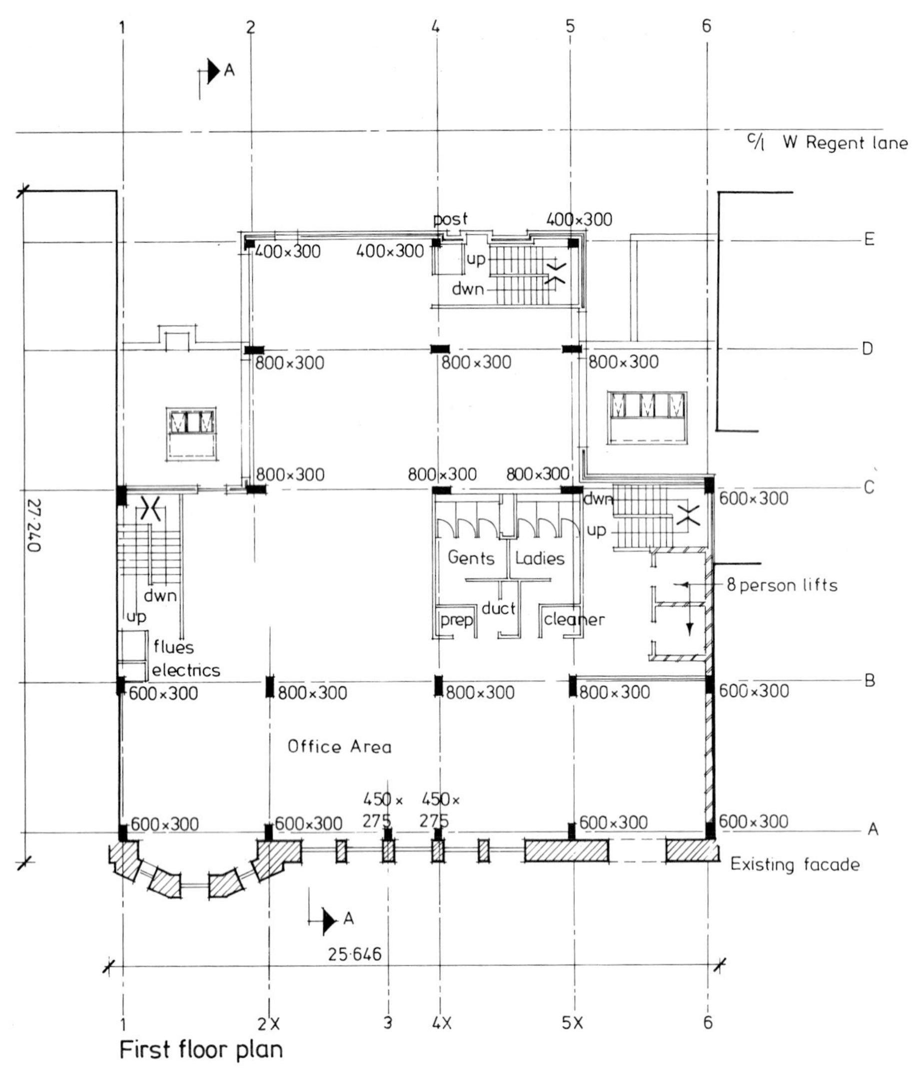

Note :- Columns 2X/A, 2X/B, 5X/A, 5X/B:- 305×165 UB steel stanchions used initially for temporary support system to facade and finally encased in concrete to form part of structural frame

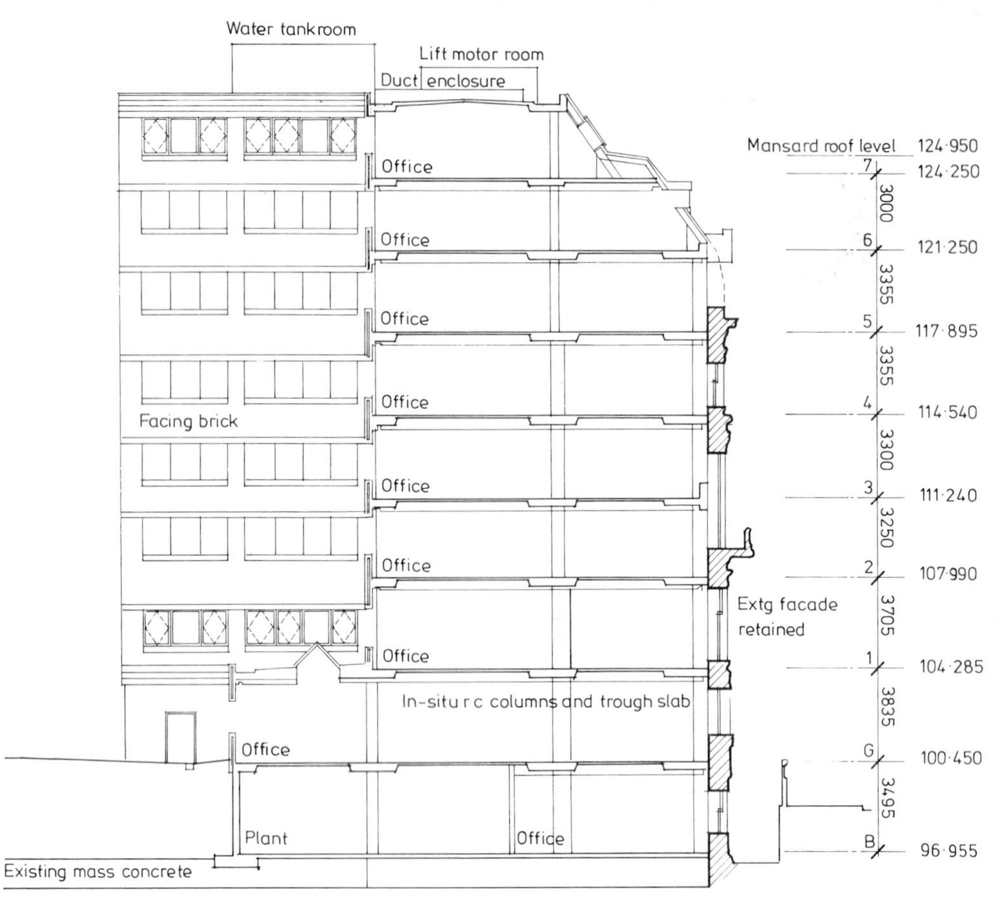

Section A–A

Figure 7.43 144 West George Street, Glasgow: first floor plan and vertical cross-section.

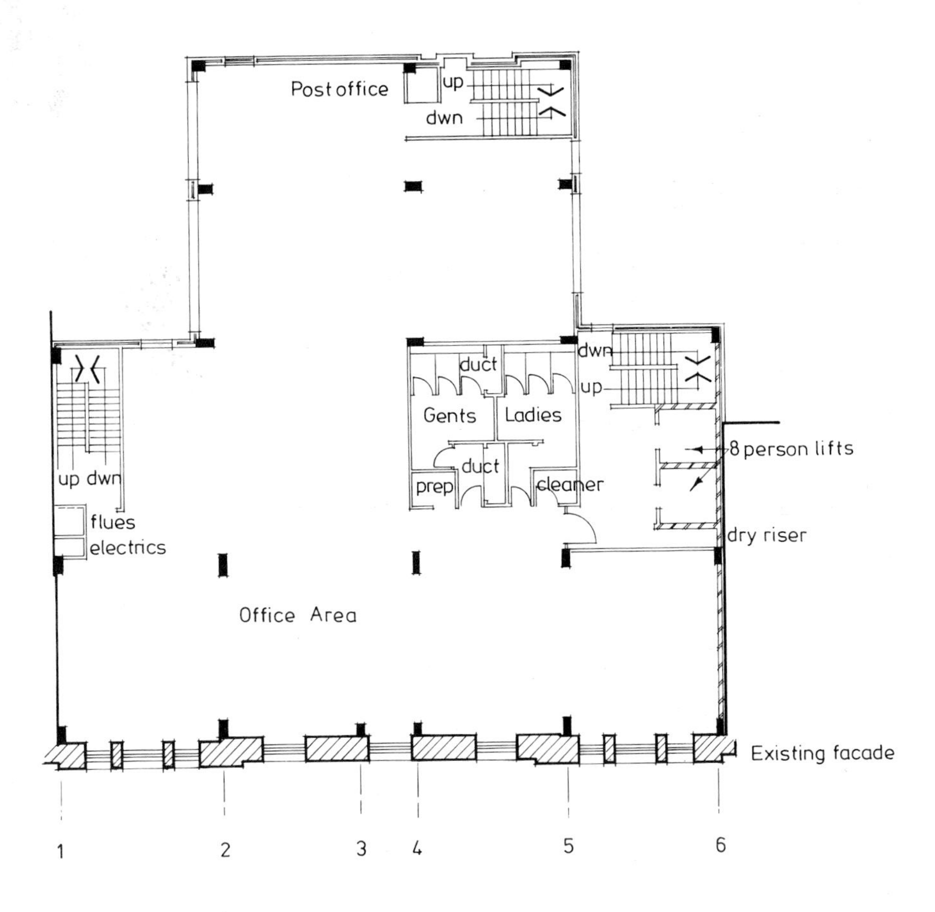

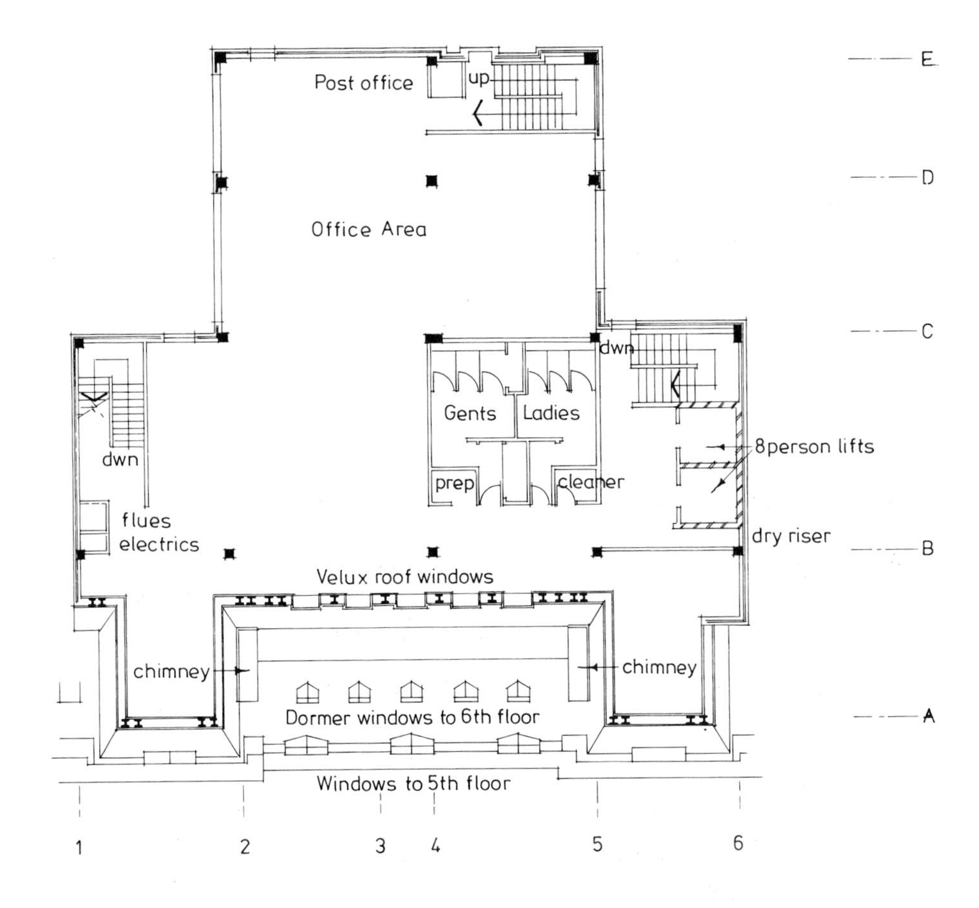

Figure 7.44 144 West George Street, Glasgow: third and seventh floor plans.

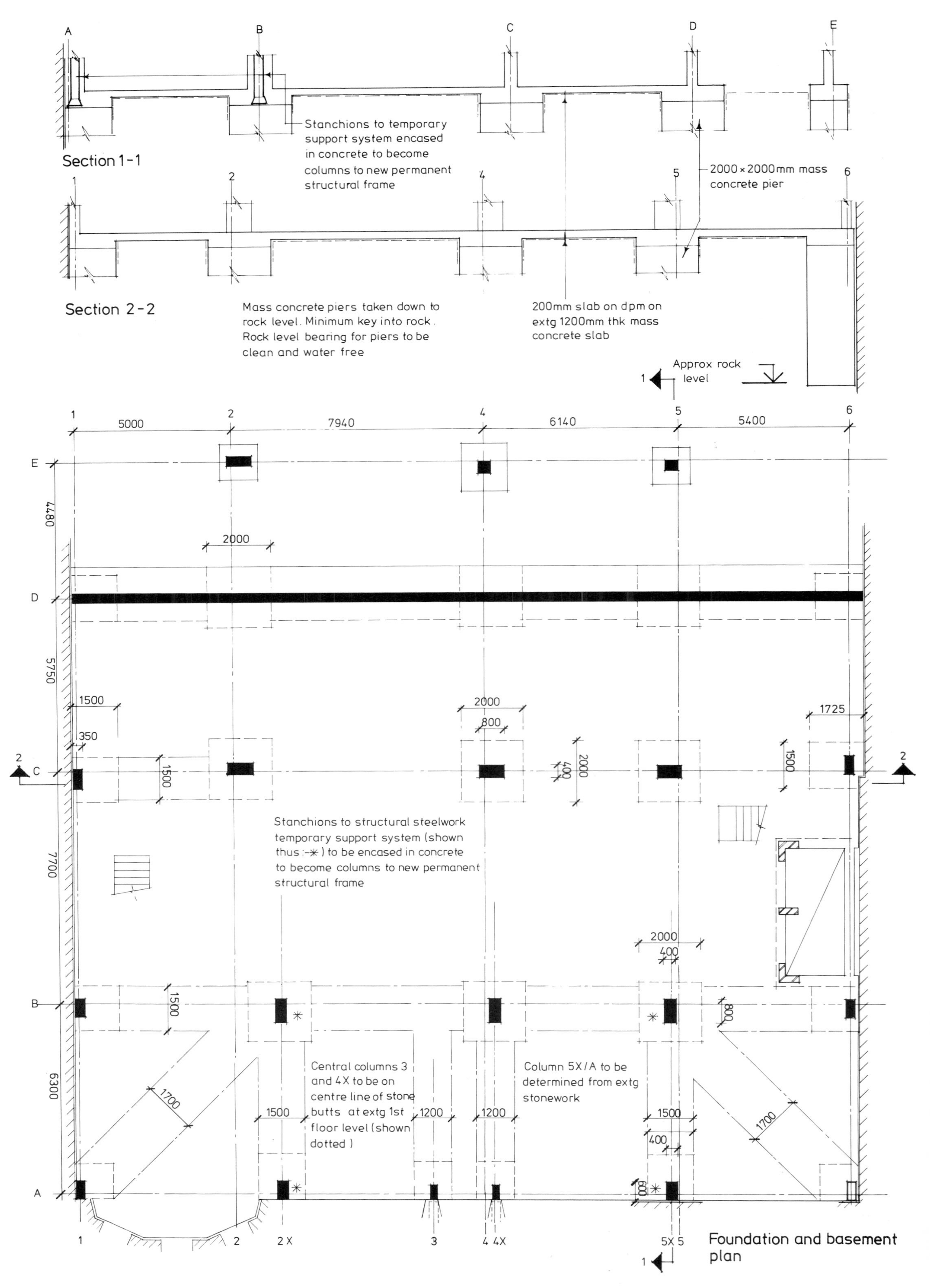

Figure 7.45 144 West George Street, Glasgow: basement plan and foundations.

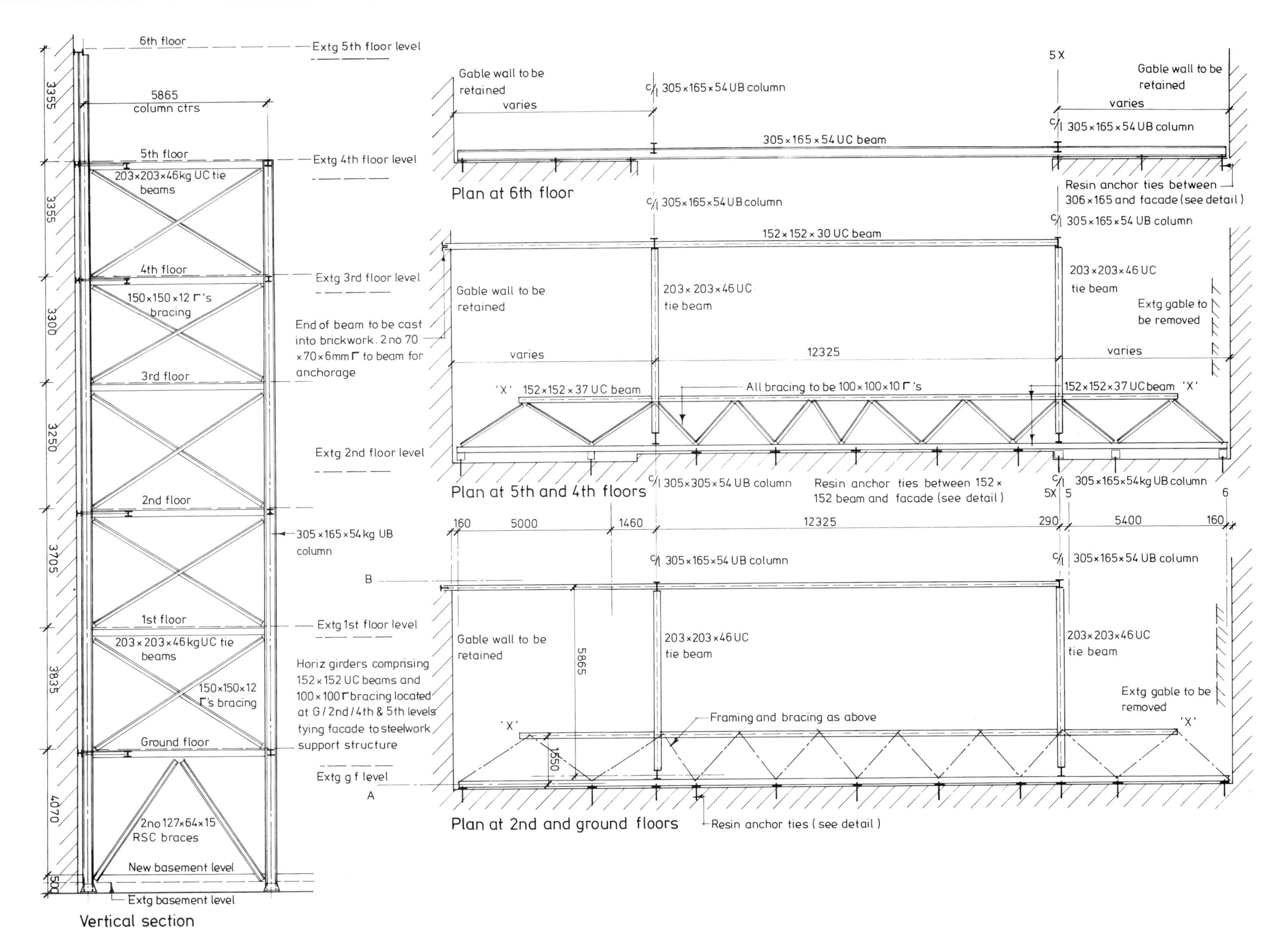

Figure 7.46 144 West George Street, Glasgow: temporary support system.

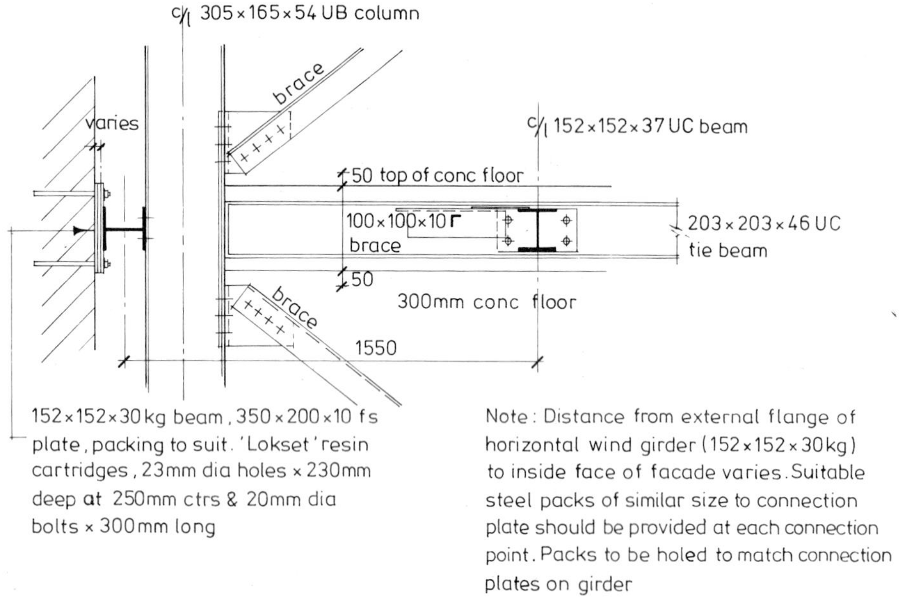

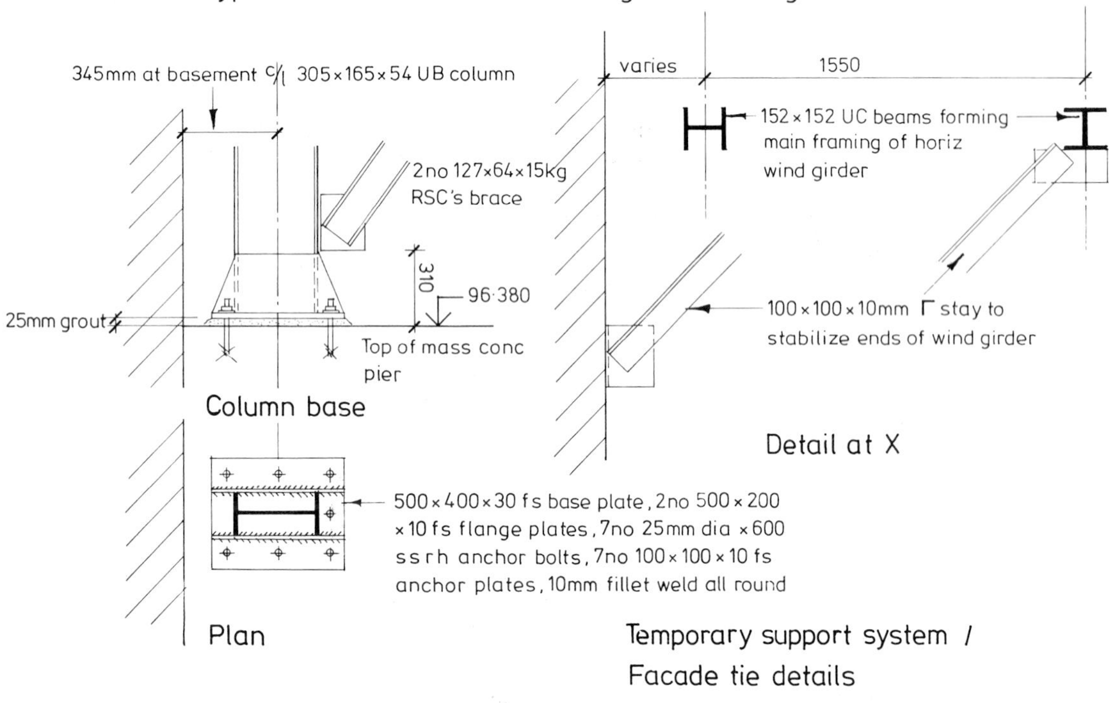

Figure 7.47 144 West George Street, Glasgow: temporary support details.

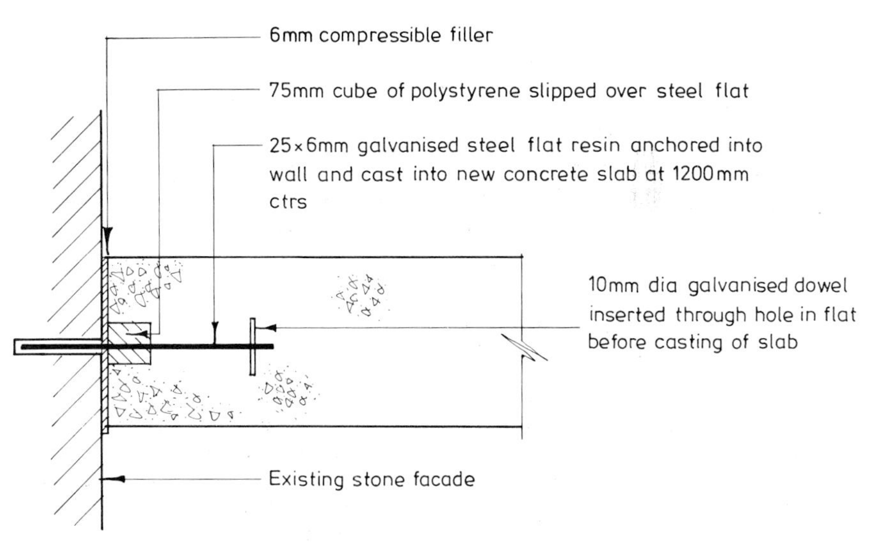

Front facade tie detail occuring at levels where no wind bracing used

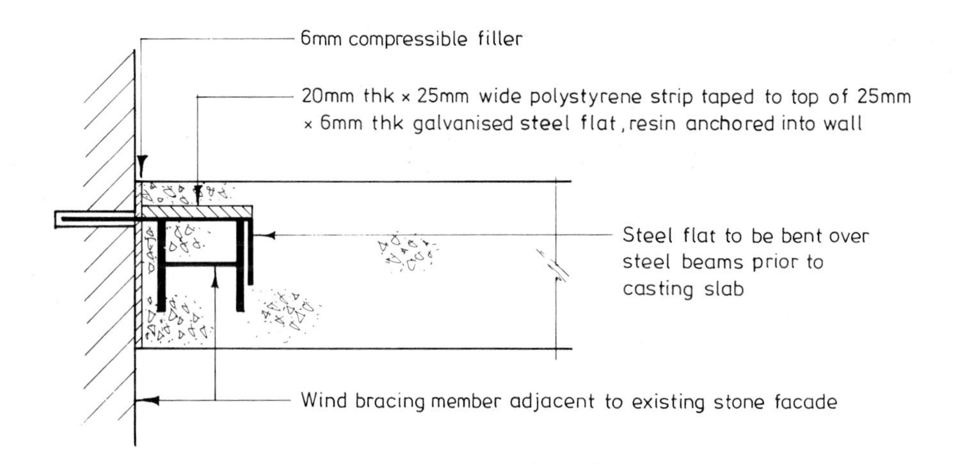

Front facade tie detail occuring at points where bracing used ie levels G, 2nd, 4th, 5th and 6th

Note: Lokset 'P' (pumped) resin used in both details

Figure 7.48 144 West George Street, Glasgow: facade-ties.

Figure 7.49 144 West George Street, Glasgow: front elevation near completion of the scheme.

Figure 7.50 144 West George Street, Glasgow: front elevation after completion of interior demolition.

Figure 7.51 144 West George Street, Glasgow: view from rear showing temporary support system.

Figure 7.52 144 West George Street, Glasgow: view from rear showing temporary support system and construction of new structure.

Figure 7.53 144 West George Street, Glasgow: view from rear showing temporary support system.

Figure 7.54 144 West George Street, Glasgow: close-up view showing resin-anchored facade-ties.

Figure 7.55 144 West George Street, Glasgow: wind-bracing member and formwork prior to casting of new concrete floor slab.

Case study five

1–3 Baxter's Place, Edinburgh

Background to the scheme

The existing Category B listed buildings formed part of what was originally a symmetrical group from 1–8 Baxter's Place, built by John Baxter in 1800. The block comprised three storeys with attics and basements, and numbers 1–2 and 6–8, the pavilion blocks, had Roman Doric pilasters with broken entablatures. The original front elevation was of rusticated ashlar to the basement, vee-jointed ashlar to the ground floor, and polished ashlar above. Numbers 1–3, like the rest of the group, were originally residential, but, from the mid-nineteenth century, commercial uses intruded. Number 3 was still partially residential until as recently as 1938, but by 1957 the lower floors of all three buildings had been converted into a retail shop. The various changes of use resulted in extensive alterations which were totally unsympathetic to the original design and badly marred the buildings' appearance. These included projecting shopfronts at ground level, and the replacement of the original roof at number 1 with a mansard. The interior of number 1 had been drastically altered on all floors, leaving none of its original internal fabric intact. Number 2 and 3 had stood empty and neglected for a long period and, although some of the original finishes had survived, they were in a very poor state of repair. Figure 7.63 shows the front of numbers 1–3 Baxter's Place prior to the commencement of the scheme.

The scheme involved complete demolition of the interior and roof of numbers 1–3, and the erection of a new steel framed structure within the original external envelope which underwent extensive restoration. In addition, the dilapidated buildings to the rear were demolished and replaced by a new steel framed building, connected to the retained building by a glazed entrance link. The restoration of the external envelope included the rebuilding in natural stone of the ground floor frontage in a style similar to the original; and the replacement of the mansard roof with a new slated roof, hipped to match the pavilion type roof at numbers 6–8. Additional features which played an important role in the success of the restoration included the provision of cast iron railings and the construction of chimney stacks to match those which had survived in the other buildings of the group. Figures 7.64 and 7.65 show the front elevations of the completed building and its group. The new framed structure erected within the retained and restored envelope is of structural steelwork founded on concrete pad foundations and supporting precast concrete floors of pre-stressed planks and filler blocks with an *in situ* concrete topping. Figures 7.56, 7.57 and 7.58 show typical floor plans and a cross section of the completed scheme.

Temporary support system

The part external/part internal temporary support system, shown in Figures 7.59, 7.60, 7.66 and 7.67, was constructed entirely from tubular scaffolding components. The use of external support elements was possible at the front of the buildings due to the particularly wide pavement, and at the rear where the ground belonged to the property and therefore did not affect any public right of way.

The front and rear elevations of the retained envelope were supported by a fully integrated system (shown in Figure 7.59) consisting of the following principal elements.

1. An external independent tied scaffold and eight sets of raking shores at the front of the building. The lower thrusts of the rakers were taken by two temporary mass concrete kentledges positioned 7.6 m and 7.7 m out from the facade.
2. An external independent tied scaffold at the rear of the building.
3. A total of six internal tubular scaffolding flying shores, in pairs at three different levels, spanning between the front and rear elevations. The ends of the flying shores were taken through the existing window openings and tied to the external shoring systems in order to complete the unified support structure.

The front and rear elevations of the retained facade were secured to this support structure using a 'collaring' technique, together with reveal ties as shown in Figure 7.60. With the former technique, each elevation was collared between horizontal scaffold tubes fixed along its inner and outer faces, the tubes in turn being carried by the main support structure. The wall was then rigidly secured between this collar of horizontal tubes using timber folding wedges packed between the tubes and the masonry. The reveal ties comprised vertical and horizontal scaffold tubes, tied to the main support structure and rigidly braced into the reveals of selected window openings using adjustable reveal pins. These reveal ties fulfilled a further important function in reducing the possibility of any adverse distortion of the window openings and associated localized movement of the masonry.

The side elevation of the retained envelope was supported by an external independent tied scaffold to which the wall was secured using tubes taken directly through holes formed in the masonry, together with an arrangement of tubing and folding wedges designed to collar the wall in the vicinity of the window openings.

Structurally, the three independent tied scaffolds to the front, rear and side elevations each comprised a series of horizontal trusses, formed by means of heavy plan bracing as indicated in Figure 7.59. As with all part-internal temporary support systems, its installation had to be properly co-ordinated with the demolition operations in order to minimize risk to the facade, and this was achieved by commencing at the top of the building and working downwards. Following the erection of the external support elements, the roof and third floor were demolished, followed by the erection of shores A.6 and B.6 (Figure 7.59) using the existing second floor as a working platform. When these two shores were in position and performing their supporting function, the existing second floor was demolished. Shores A.5 and B.5 were then erected and the existing first floor subsequently demolished. Finally, shores A.4 and B.4 were installed, followed by demolition of the existing ground and basement floors.

A further important consideration in locating the internal flying shores was ensuring that they would not interfere or conflict with the construction of the new structure, since the shores had to remain in position, performing their supportive function, until the new frame and floors had been constructed and the facade permanently tied back to them.

Facade ties

The retained walls of 1–3 Baxter's Place were tied back to the edges of the new concrete floor slabs by means of resin-anchored ties at four levels. Six different tie designs were necessary to suit the varying conditions at the junction of the new and existing structures, and their locations are shown in Figure 7.61. 'Kemfix' masonry anchors were used to form the facade ties which are detailed in Figure 7.62 and are described below.

Type 'A'

The majority of resin-anchored facade ties employ tie bars anchored directly into resinous mortar inside pre-drilled holes in the masonry. The type 'A' tie is different in that the tie bar is inserted into an internally threaded steel socket, itself previously resin-anchored into the masonry. The type 'A' tie was used where no structural steel frame members were present in the edges of the new floor slabs at their junction with the retained facade, and the procedure for their installation was as follows.

1. 18 mm diameter × 100 mm deep hole drilled into masonry using a rotary percussion drill.
2. 'Kemfix' resin capsule inserted into hole. The capsule comprises a sealed glass tube containing polyester resin, quartz granules and a phial of hardener which, when the capsule is broken and they are mixed, form a rapid-setting resinous mortar.
3. 100 mm long internally threaded stainless steel 'Kemfix' socket spun into the hole using an hexagonal driver bar attachment. The purpose of this operation was to break the resin capsule and mix its constituents to form the rapid-setting resinous mortar which anchored the socket into the masonry.
4. 12 mm diameter × 350 mm long mild steel tie bar with threaded end inserted into the internally threaded stainless steel socket, leaving approximately 250 mm of the tie bar projecting into the zone of the new floor slab.
5. *In situ* concrete to edge of new floor slab cast around the projecting tie bar.

Type 'B'

Type 'B' facade ties were used where universal section beams of the new steel frame were present in the edges of the new floor slabs at their junction with the retained facade. The unevenness of the facade made it necessary to devise two different designs for the type 'B' ties in order to cater for the varying gap between the new steel beams and the inner face of the facade. One design was employed where the gap was approximately 100 mm; the alternative design being used where the gap exceeded 100 mm.

The tie designed to suit a gap of approximately 100 mm between the new beam and the facade comprised a 200 × 150 × 12 mm mild steel angle, 80 mm long, resin-anchored to the facade and welded to the upper flange of the new steel beam. The angle was fixed to the facade using a 16 mm diameter 'Kemfix' stud resin-anchored directly into an 18 mm diameter × 152 mm deep hole previously drilled into the masonry. After installation of the tie, the *in situ* concrete to the new floor slab was cast, partially encasing the mild steel angle together with the new beam to which it had been welded.

The alternative design for the type 'B' facade tie, used where the gap between the new steel beam and the masonry exceeded 100 mm, comprised an 800 mm long × 12 mm diameter mild steel tie bar anchored into the masonry using a 'Kemfix' internally threaded socket, similar to that used for the type 'A' facade tie. Following its anchoring into the masonry, the tie bar was bent around the upper flange of the new steel beam before being fully encased, together with the beam, when the *in situ* concrete to the new floor slab was cast.

Type 'C'

Type 'C' facade ties were used where RSJ section beams of the new steel frame were present in the edges of the new floor slabs at their junction with the retained facade. In common with the type 'B' ties, two different designs were necessary to suit the varying gap between the new steel beams and the uneven masonry of the facade.

The first design, used where the gap was approximately 150 mm, comprised a 200 × 150 × 12 mm mild steel angle, 80 mm long, resin-anchored to the facade and welded to the upper flange of the new RSJ. This tie was identical in design to the type 'B' tie, described earlier.

The alternative design for the type 'C' facade tie, used where the gap between the steelwork and facade exceeded 150 mm, consisted of a 650 mm long × 12 mm diameter mild steel tie bar anchored into the masonry using a 'Kemfix' internally threaded socket. After being anchored into the masonry, the tie bar passed under the RSJs bottom flange and was then bent upwards to achieve an anchorage in the *in situ* concrete floor slab edge, within which both the tie bar and the RSJ were fully encased.

Type 'D'

Type 'D' facade ties occurred in the side elevation where the new brickwork lift shaft walls joined the existing facade. The ties comprised 1600 mm long × 12 mm diameter mild steel bars, anchored into the facade masonry using 'Kemfix' internally threaded sockets, and built into the mortar joints of the new lift shaft walls.

Differential settlement

The data obtained from subsoil investigations beneath the existing building led to the conclusion that the new structure would undergo negligible settlement if adequate foundations were designed to carry its new steel stanchions. The foundations are shown dotted in Figure 7.56 and are described in detail later. The knowledge that settlement would be negligible therefore enabled the facade ties and the interface detail between the new and existing structures to be designed without any provision for differential movement. The details in Figure 7.62 show that each facade tie produced a fully rigid connection, and the interface between the new and existing structures did not incorporate the slip surface that would have been essential had a significant degree of differential movement been predicted.

A further factor affecting provision for differential settlement was that the unevenness of the inside face of the retained walls ruled out the incorporation of an effective vertical slip plane between the edges of the new floor slabs and the existing masonry. This, together with the fact that the new structure's settlement would be negligible, therefore made any provision of the slip surface at the interface both impractical and unnecessary.

Foundation design

The layout of the new structural steel frame required that ten stanchions be located immediately adjacent to the retained walls of 1–3 Baxter's Place. The necessary proximity of the new stanchions to the retained facade, and the fact that the facade's foundations were very shallow, led to a number of problems in the design of the new foundations. The principal problem was to minimize any disturbance of the existing shallow foundations which could have resulted in collapse of the facade, and the method devised to overcome it, shown in Figure 7.56, involved using two basic foundation types.

1. The three new stanchions located in the corners of the retained envelope were founded on balanced-base foundations which obviated the need to undermine the existing foundations (section 6.5). The eccentric loads on the foundations of the north-east and south-east corner stanchions were balanced by structurally linking them to the base of the internal stanchion at grid position $A_1$2. Similarly, the foundation of the north-west corner stanchion was structurally linked to that of the internal stanchion at grid position $E_2$2.
2. The use of balanced-base foundations for the remaining seven stanchions proved to be impractical and therefore undermining of the existing foundations was necessary. However, the extent of the undermining was kept to a minimum by designing unusually narrow pad foundations; the narrow dimension limiting the required amount of undermining. Initially, these foundations were designed to have widths of only 1 m, but this would have required their lengths to be excessive, and ultimately a width of 1.5 m was used, which enabled their construction to be successfully completed with minimal disturbance to the retained facade.

Architects Robert Hurd & Partners, Edinburgh.

Consulting engineers Laing Properties, Manchester.

Main contractor John Laing Construction.

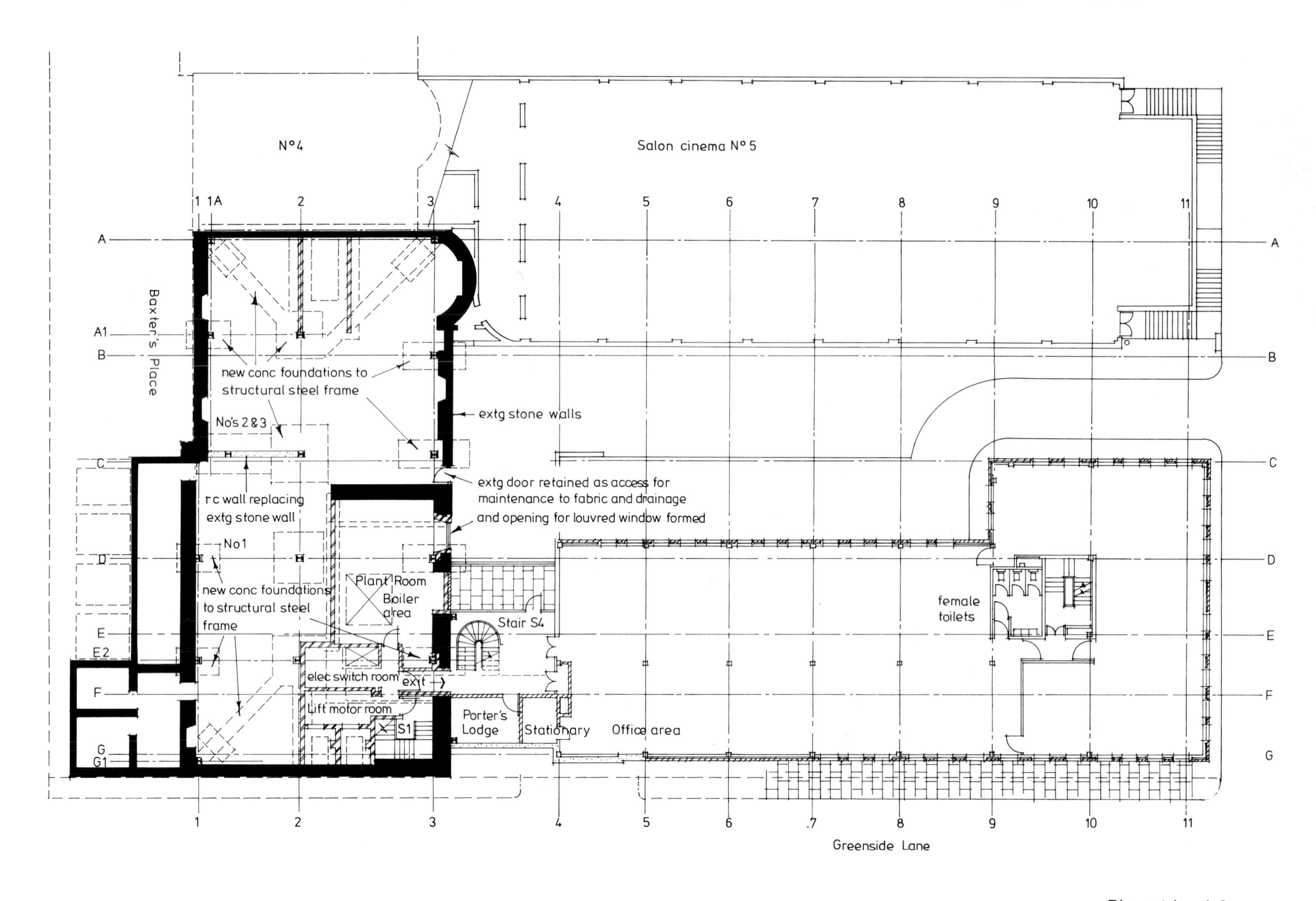

Figure 7.56 1–3, Baxter's Place, Edinburgh: level 2 floor plan.

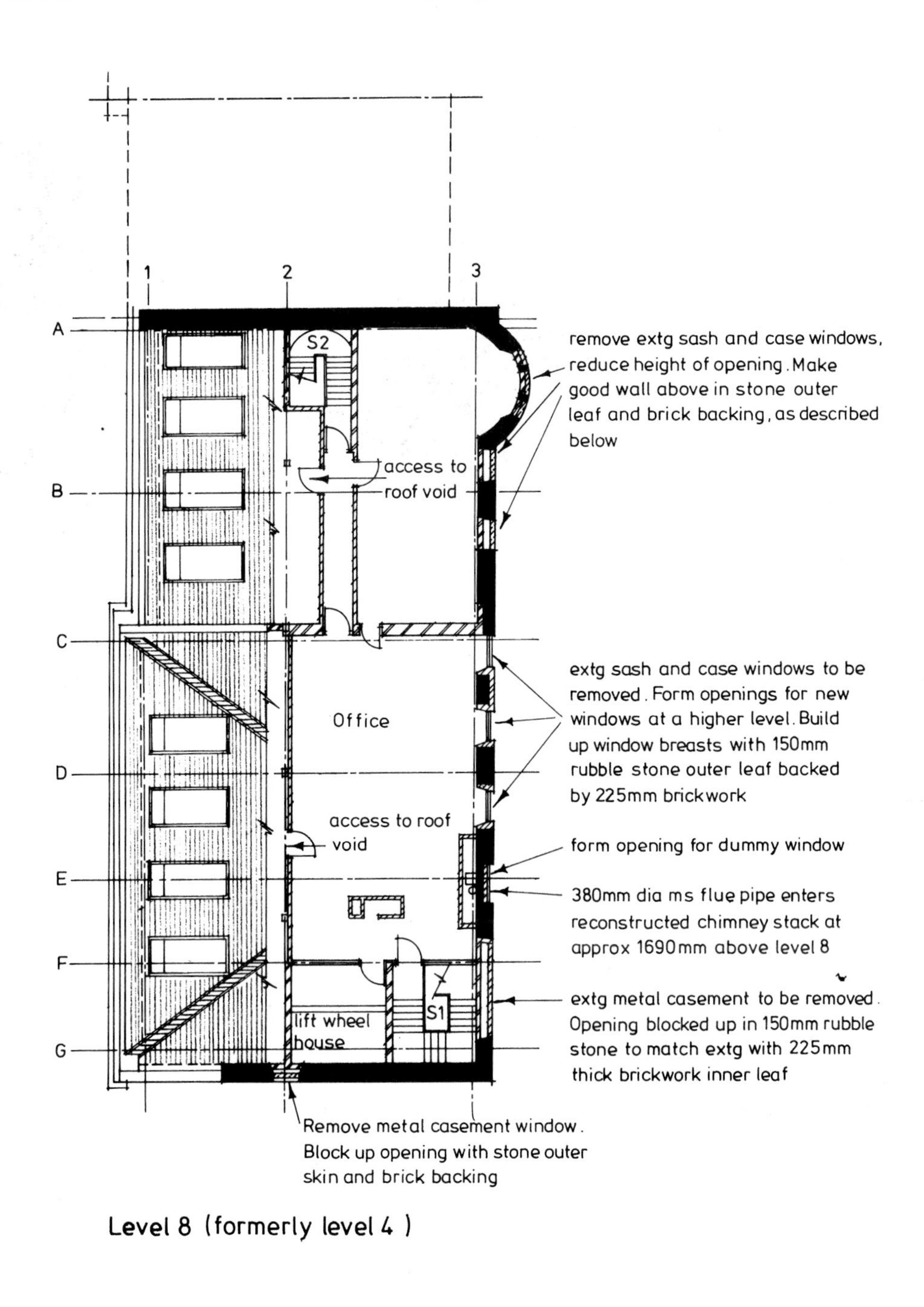

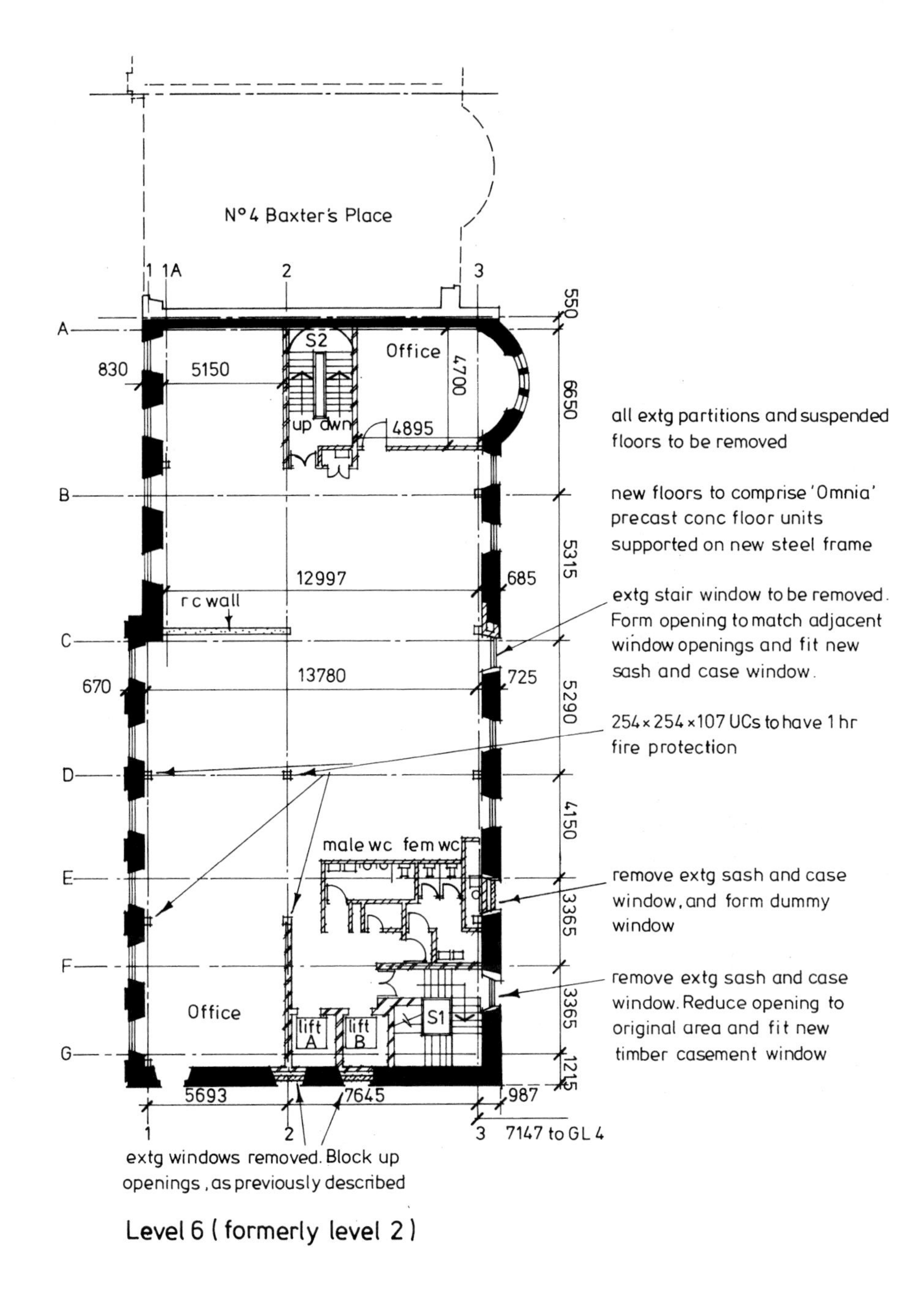

Figure 7.57 1–3, Baxter's Place, Edinburgh: levels 6 and 8 floor plans.

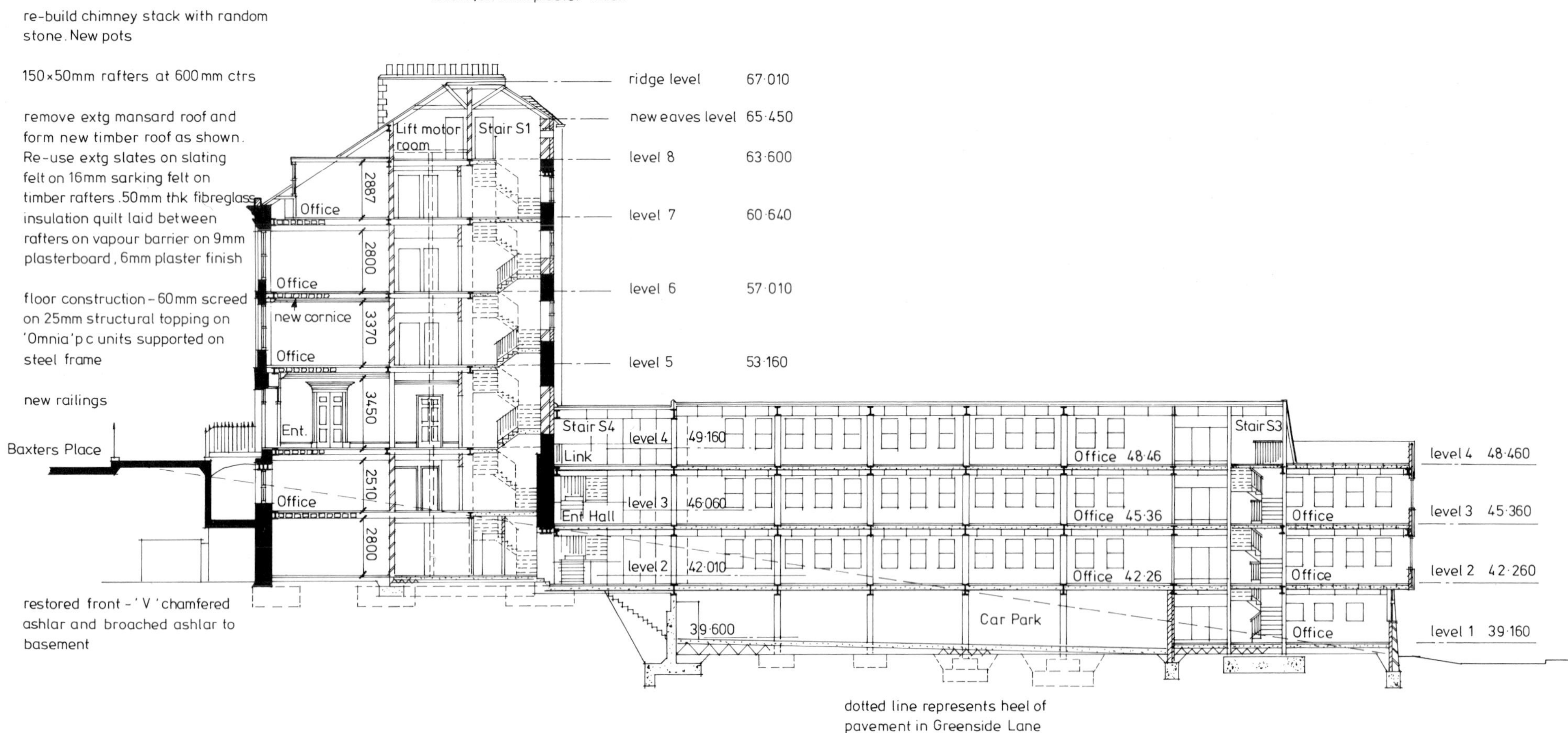

Figure 7.58 1–3, Baxter's Place, Edinburgh: longitudinal cross-section.

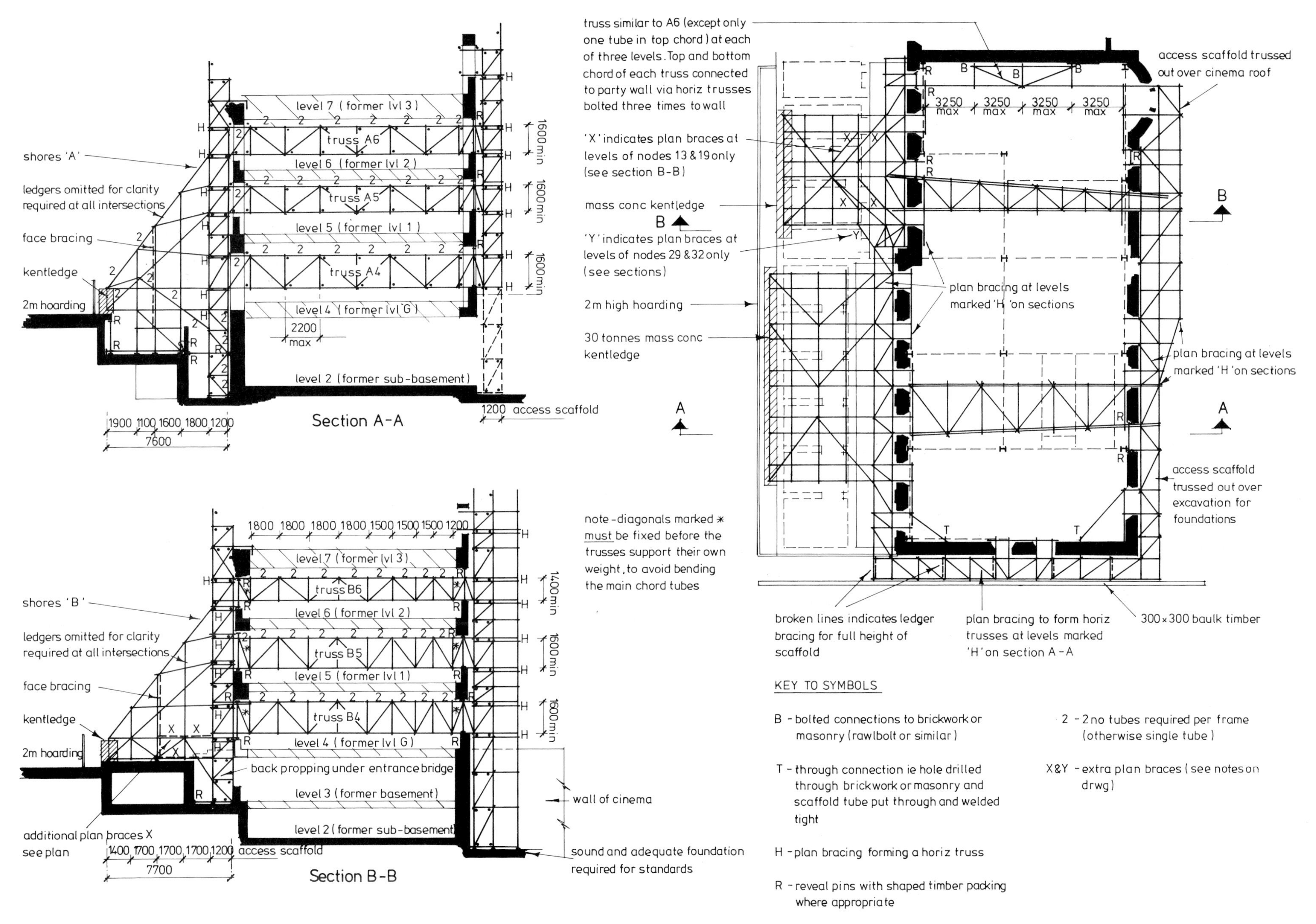

Figure 7.59 1–3, Baxter's Place, Edinburgh: temporary support system.

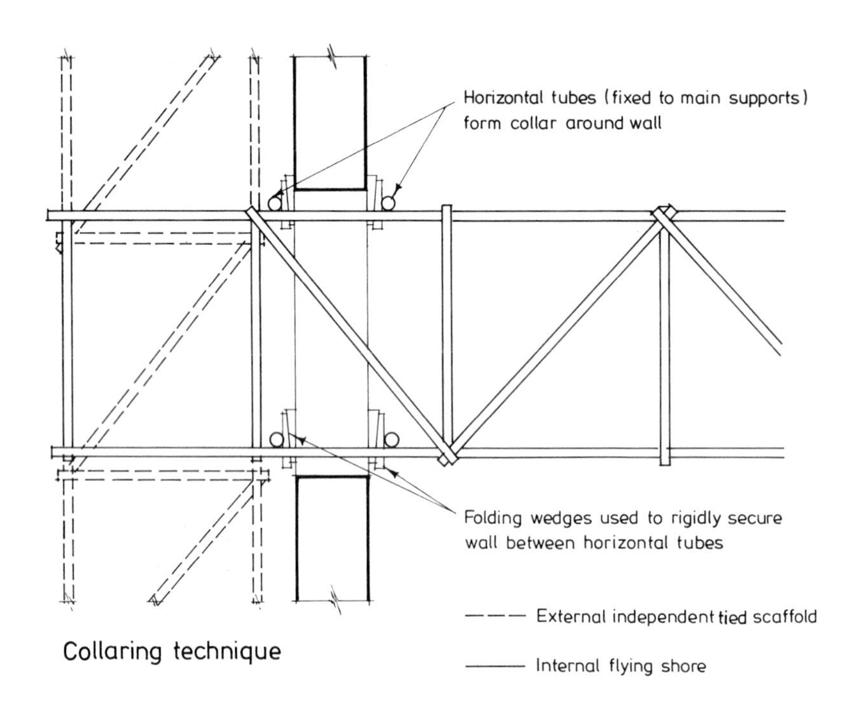

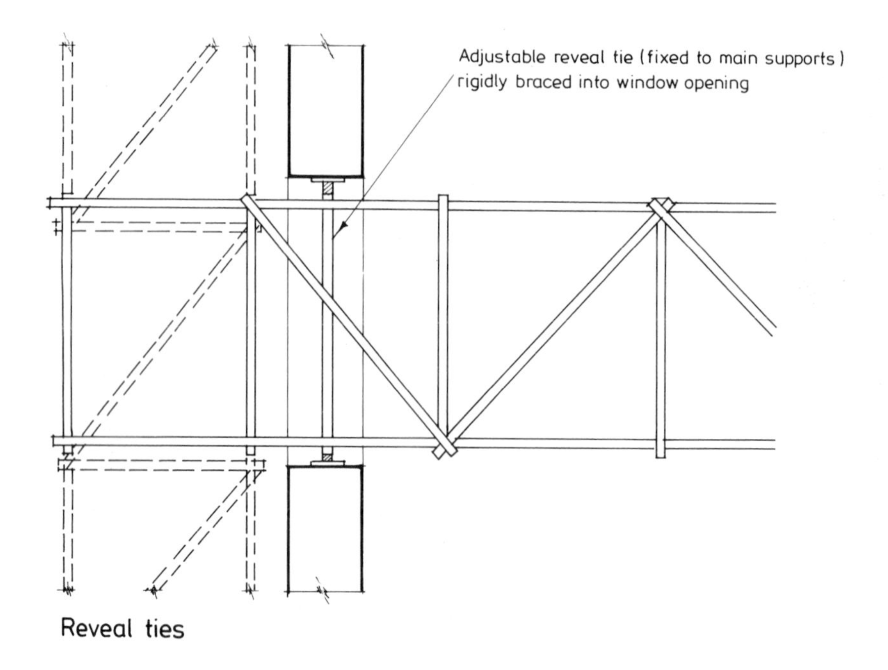

Figure 7.60 1–3, Baxter's Place, Edinburgh: temporary support details.

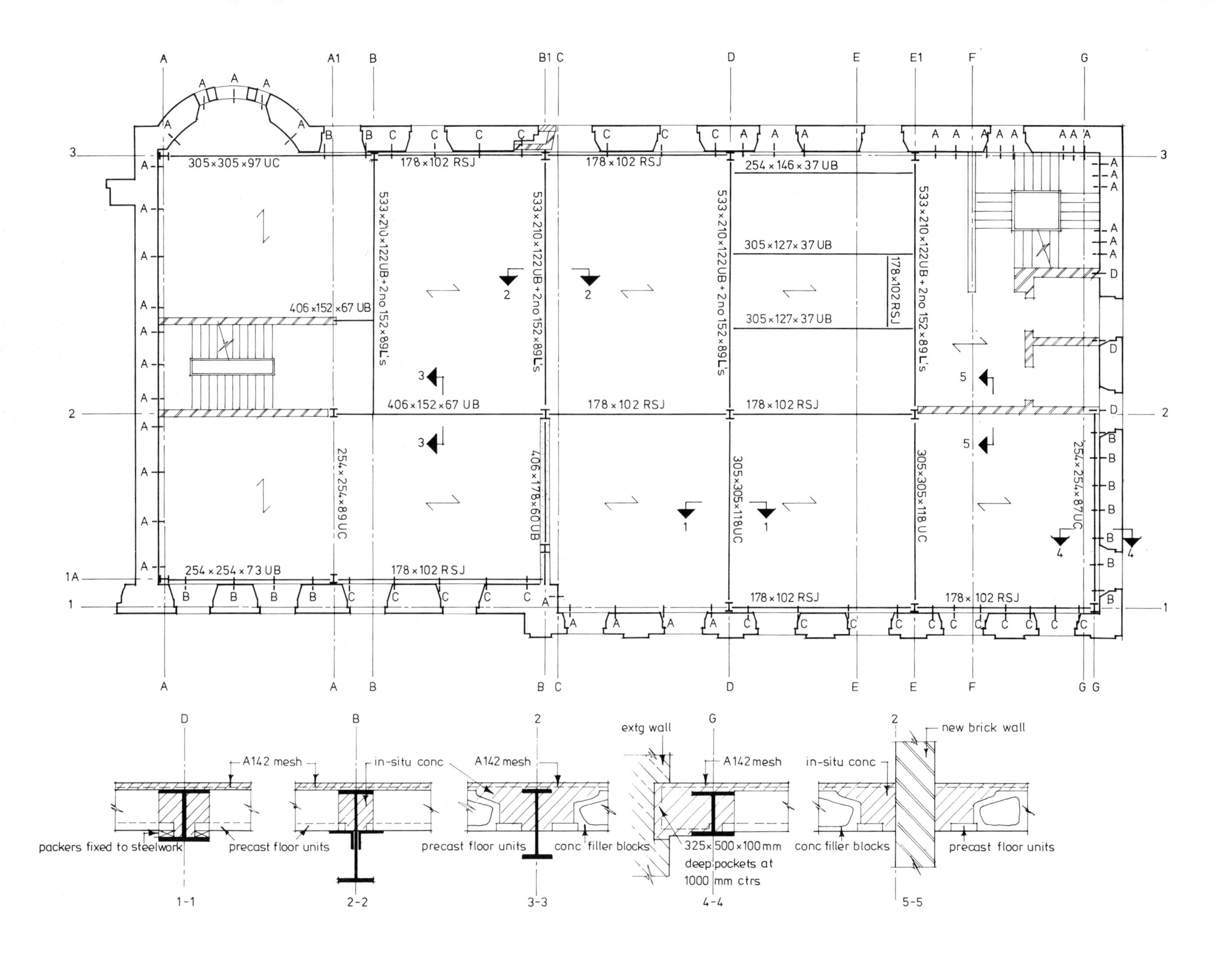

Figure 7.61 1–3, Baxter's Place, Edinburgh: typical floor plan showing new structure and facade-ties.

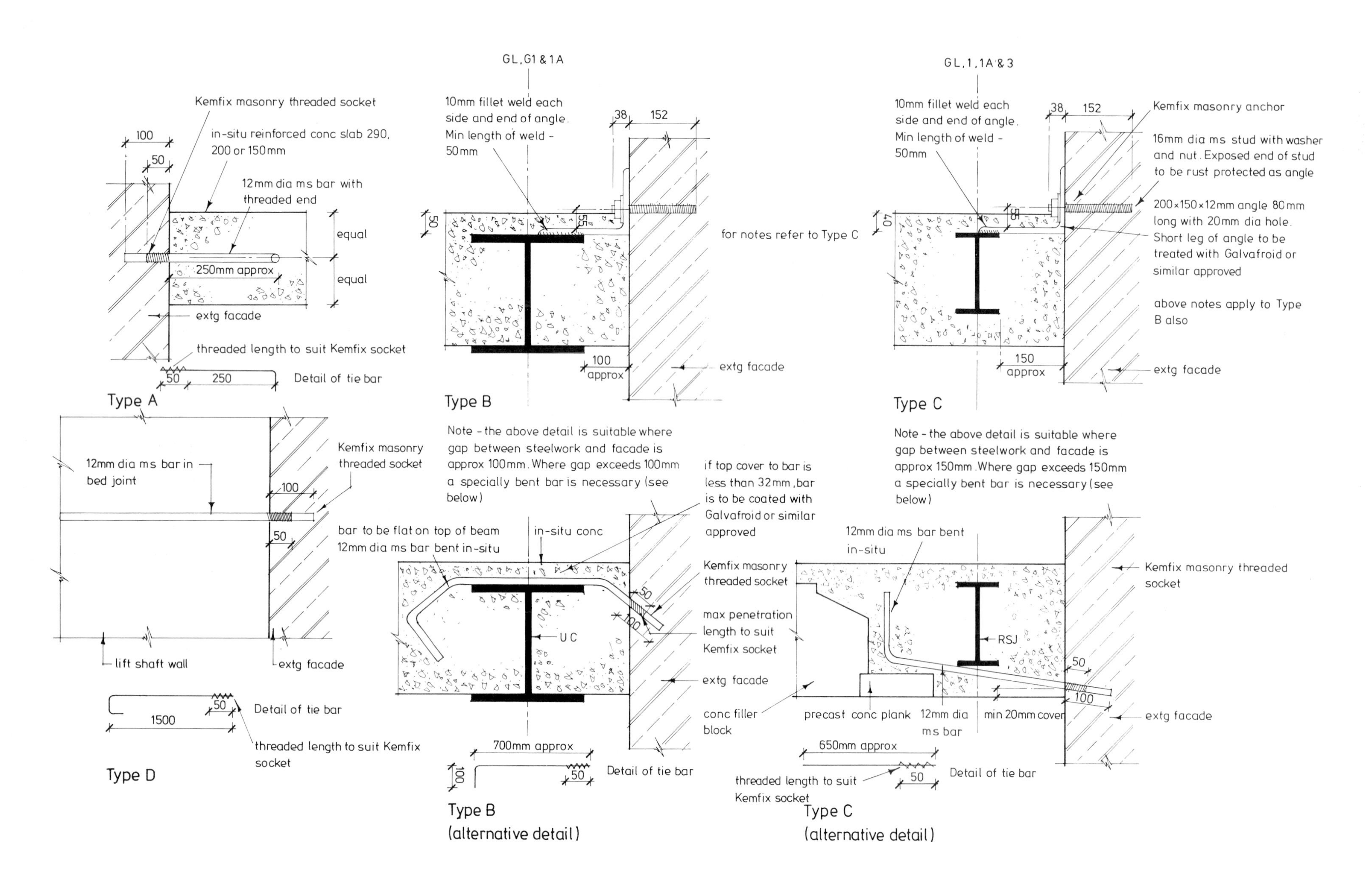

Figure 7.62 *1–3, Baxter's Place, Edinburgh: facade-tie details.*

Figure 7.63 1–3, Baxter's Place, Edinburgh: front elevation before redevelopment.

Figure 7.64 1–3, Baxter's Place, Edinburgh: front elevation after re-development.

Figure 7.65 1–3, Baxter's Place, Edinburgh: front elevation after re-development.

Figure 7.66 1–3, Baxter's Place, Edinburgh: temporary support system to front elevation.

Figure 7.67 1–3, Baxter's Place, Edinburgh: interior view after completion of demolition works showing temporary flying shores.

Figure 7.68 1–3, Baxter's Place, Edinburgh: interior view showing new steel frame erected within the retained envelope.

Case study six

3–13 George Street, Edinburgh

Background to the scheme

The development at 3–13 George Street affected a group of three buildings consisting of a Category A listed building (numbers 3–7) at the eastern end, an unlisted 1930s building (numbers 9–11) in the centre, and a Category C listed building (number 13) at the western end. Figure 7.76 shows the group viewed from the western end before work commenced.

The development, which provided modern office accommodation over the whole site, involved retaining the Category A listed building in its entirety, subjecting it to only minor internal alterations. This building, designed in the Palladian style by J.M. Dick Peddie and built in 1897, possessed many valuable interior as well as exterior features, which, in addition to the suitability of its internal layout and structure, led to its complete retention. The Category C listed building, number 13 George Street, designed by W. Hamilton Beattie and built in 1898, had no interior features of significant merit, its listing being due mainly to the value of its facade. The local planning authority therefore granted permission for the total demolition of its interior and the retention of its facade only, the interior being replaced by a new steel framed structure connected to that which replaced the totally demolished unlisted building in the centre of the group. Figure 7.77 shows the front and west elevations of the completed scheme.

One of the reasons for the development was to overcome the problems created by the different floor levels of the three buildings which, despite their separate exterior identities in the new scheme, were to continue to function as a single office building. The gutting of number 13 and the total demolition of the centre building overcame this problem by allowing the insertion of new floors at the same level which linked with the retained floors of numbers 3–7.

The new structure erected within the original shell of number 13 George Street comprises a structural steel frame supporting pre-cast concrete floors. Most of the frame's beams were built into, and supported by, the retained facade, the connections being made using concrete padstones. These connections ensured that the facade was tied back to the new structure and, in addition, enabled the facade to continue its load-bearing function and carry some of the new building's loads. This use of a retained facade to support the new structure is most unusual and can only be considered where the structural integrity of the facade can be maintained. In this scheme, virtually the whole envelope of the existing building was retained, giving it sufficient stability and integrity to enable it to continue as a load-bearing element for the new structure. The design for the new structure enabled the total number of floors (including basement and attic) to be increased from the original six, to seven. The addition of the extra floor, which was located between the original ground and first floor levels, required further major constructional operations, including lowering the original basement level by almost half a metre, together with the underpinning of the retained facade's existing foundations.

Temporary support system

The retained shell of number 13 George Street was supported by a part external/part internal system constructed mainly from standard tubular scaffolding components. External supports to all the elevations were possible due to the ample amount of clear space around the building, which included spacious pedestrian thoroughfares at the side and rear, and a wide pavement adjoining George Street at the front. Figures 7.72–7.74, 7.78 and 7.79 show the design and layout of the temporary support system, which comprised the following principal elements.

1. A heavily braced independent scaffold around the outside of the retained shell, to which the north, west and south walls were secured by means of through ties, horizontal tubes, timber wall plates and folding wedges (Figure 7.72). The through ties, taken from the independent scaffold through windows and other openings formed in the masonry, supported horizontal tubes on each side of the facade between which it was rigidly collared by means of timber wall plates and folding wedges.
2. A unit beam scaffold erected outside the windowless east wall which was secured to it by means of ties passing through holes formed in the masonry, horizontal tubes, timber wall plates and folding wedges. Lack of space adjacent to the east wall allowed only a narrow scaffold which was unable to support the wall unaided. End restraints were therefore provided to this unit beam scaffold by securing one end to the north elevation's independent scaffold, and the other end to the four internal raking shores.
3. A total of ten horizontal rolled steel channels secured to the external faces of the north and south elevations and connected to the ends of the west elevation's independent scaffold. This arrangement, shown in Figure 7.73, was designed to provide additional stability to the independent scaffold and, at the same time, give extra support to the north and south elevations. The 152 × 89 mm rolled steel channels were secured to the walls using 25 mm diameter threaded rods passing through the bricked-up window openings in the south elevation and holes drilled through the masonry in the north elevation (detail Figure 7.73). The free ends of the channels, which projected into the ends of the west elevation's independent scaffold, were then connected to the latter by welding (detail Figure 7.73).
4. Six internal flying shores constructed from tubular steel scaffolding and connected to the external support system via window openings and other holes formed through the masonry. Four of these shores were located, one above the other, between the west and east elevations (Figure 7.72, plan and section A–A), their function being to provide additional stability and restraint to the narrow external unit beam scaffold which, together with the flying shores, supported the north elevation. All four flying shores were connected to the west elevation's external independent scaffold by extending their ends through existing window openings; the uppermost shore being positioned differently from the others to enable its connection to be made through the larger bay window. There were no window openings in the retained north elevation through which to connect the other ends of the flying shores to the external unit beam scaffold, and thus special holes had to be formed through the masonry for this purpose. Figure 7.78 shows some of these connection holes and part of the unit beam scaffold. The two remaining flying shores, which were tapered on plan, were installed at the front of the building at levels one and three, between the west elevation and the south (front) elevation. Their function was to form a structural tie between the west elevation's external scaffold and the free end of the south elevation in order to increase the overall support to the latter. The ends of these shores were connected to the west elevation's external scaffold via the existing bay window openings, and to the through ties from the south elevation's external scaffold which penetrated the bricked-up window openings.
5. An arrangement of ties, struts and bracings, shown in Figure 7.74, designed to stabilize the bay windows in the west elevation and to increase the number of ties between that elevation and the independent scaffold. Horizontal ties taken from the independent scaffold, through the middle bay window in each group, were used to support vertical members inside the bay. These vertical members, in turn, supported internal horizontal struts which, by means of adjustable endplates and plywood packings, were securely

braced between the window reveals. Each framework was completed by the addition of diagonal cross-bracing which was installed to increase its overall rigidity.

Facade ties

Two methods were used to tie back the retained facade to the new steel framed structure. The first entailed building the ends of the new steel beams into the facade by ragbolting them to mass concrete padstones previously cast into pockets formed in the masonry. These padstone connections, which are shown in Figures 7.69 and 7.75, were formed in the following manner:

1. pocket formed in inside face of facade;
2. mass concrete padstone, 450 × 220 × 220 mm, cast into bottom of pocket with two holes left for ragbolts;
3. inner end of new steel beam connected to frame, outer end positioned inside pocket and supported by padstone;
4. new steel beam secured to padstone using two 225 mm long × 20 mm diameter ragbolts through bottom flange, grouted into preformed holes;
5. pocket made good with brickwork.

In addition to acting as major structural ties securing the facade to the new steel frame, the ten padstone connections made at each floor level also enabled the facade to be used as a load-bearing element supporting parts of the new structure, but this is uncommon owing to the general instability of retained facades, which normally rules out their use as load-bearing elements.

The second method used to tie the facade back to the new structure employed resin-anchored ties made between secondary members of the new steel frame and the masonry. The facade was resin-anchored, using 'Lokset' cartridges and tie bars, to 305 × 102 mm steel channels spanning between the main floor beams, and positioned immediately adjacent to the facade's inner face. The procedure used in making these ties, which are shown in Figures 7.69 and 7.75, is described below.

1. 305 × 102 mm rolled steel channels, with 20 mm diameter holes pre-drilled through webs at 600 mm centres, are positioned as close to masonry as possible and connected to main floor beams using steel angle cleats.
2. 130 mm deep × 19 mm diameter holes are drilled into masonry at 600 mm centres (through pre-drilled holes in webs of channels).
3. CBP 'Lokset' resin cartridges, 130 mm long × 18 mm diameter, are inserted into holes in masonry.
4. 225 mm long × 16 mm diameter 'Lokset-Rebar' tie bars are spun into the resin cartridges inside holes using power tool. The operation breaks the plastic cartridge and mixes its constituents which combine to form the rapid-setting resinous mortar.
5. Resinous mortar is left to harden, firmly anchoring the tie bars into the facade masonry and leaving the threaded end of each bar protruding through its corresponding hole in the steel channel.
6. Connection is completed by securing washers and nut over the protruding threaded outer end of each resin-anchored tie bar.

The method used to tie the bay windows in the west and south elevations back to the new structure, although based on the same type of resin-anchors, was different from that described above. The bays were resin-anchored at 600 mm centres to 150 × 100 mm steel angles which were, in turn, connected to the new steel frame via the 175 mm thick *in situ* concrete floor slab in each bay. This arrangement is shown in Figure 7.69 and detailed in Figure 7.75. The detail also shows the 35 mm high × 10 mm wide mild steel lugs which were fillet welded to the steelwork to facilitate an efficient structural bond between the steel and concrete.

Differential settlement

In order to lower the basement level within the retained shell, and to enable it to carry loads from the new structure, the walls had to be underpinned with new mass concrete foundations. The underpinning, which extended approximately one metre deeper than the original foundations, inevitably resulted in the retained facade undergoing some settlement which, it was predicted, would be of a similar magnitude to, and keep pace with, the settlement of the new structure. This, together with the prediction that the settlement of the existing and new structures would be minimal, enabled the facade ties and interface details to be designed without any provision for differential settlement. The padstone connections and the resin-anchored facade ties, shown in Figure 7.75, were, therefore, both designed as fully rigid connections with no allowance for differential movement between the new and existing structures.

Foundation design

As previously stated, new foundations, in the form of mass concrete underpinning, had to be provided to the retained facade to enable it to carry the new structure's loads and allow the original basement level to be lowered. The underpinning, shown with the new structure's foundations in Figures 7.70 and 7.71, was executed in separate 'legs' which were excavated, concreted and 'pinned-up' with semi-dry, hand packed concrete in lengths not exceeding 1200 mm. This ensured that the unsupported length of wall at any one time was kept to a minimum, thereby reducing the risk to the retained facade.

The use of the underpinned facade as a major load-bearing element meant that only a small number of additional foundations were necessary to complete the support of the new structure. These foundations, to the stairwell, services duct and stanchions, were designed so that their settlement would be compatible with that of the underpinned facade, thereby minimizing the possibility of any distortion of the new frame to the detriment of the combined new and existing structures.

Client Standard Life Assurance Company, Edinburgh.

Architects Michael Laird & Partners, Edinburgh.

Quantity surveyors John Dansken & Purdie, Edinburgh.

Consulting engineers Blyth & Blyth Associates, Edinburgh.

M. & E. Engineers Mitchell Dey Norton & Partners, Edinburgh.

Main contractor Melville, Dundas & Whitson, Edinburgh.

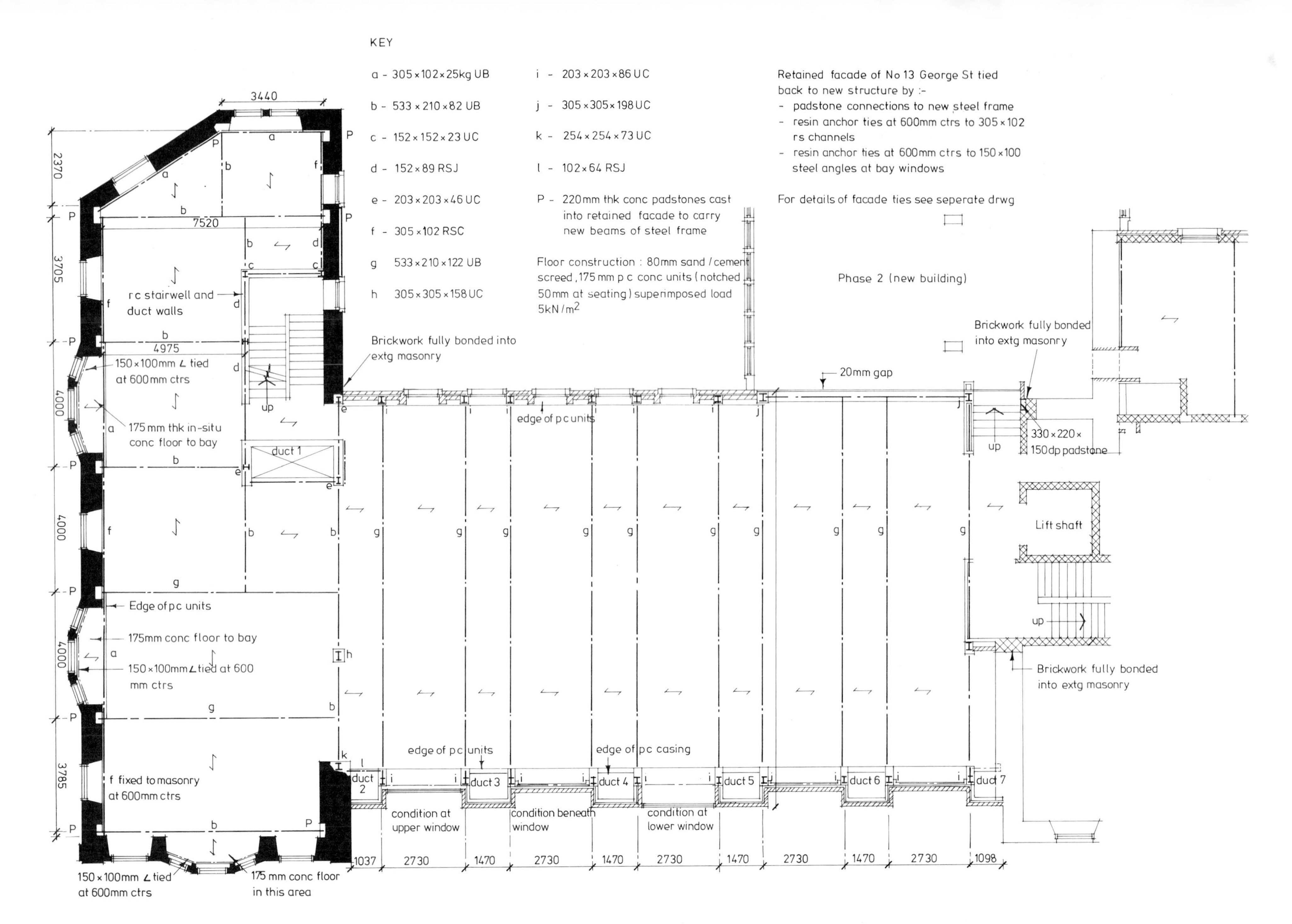

Figure 7.69 3–13, George Street, Edinburgh: first to fourth floor plans showing locations of facade-ties.

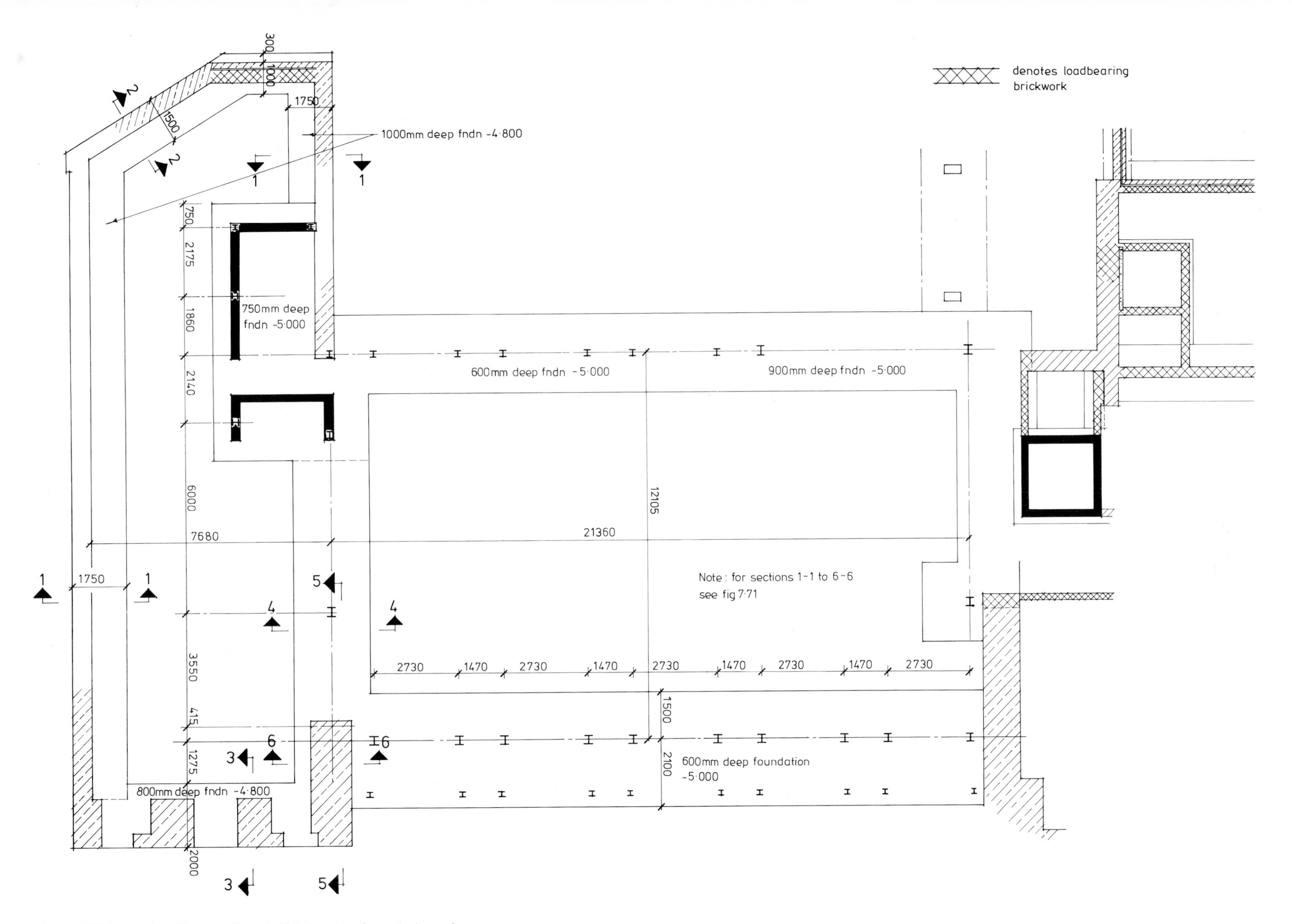

Figure 7.70 3–13, George Street, Edinburgh: foundation plan.

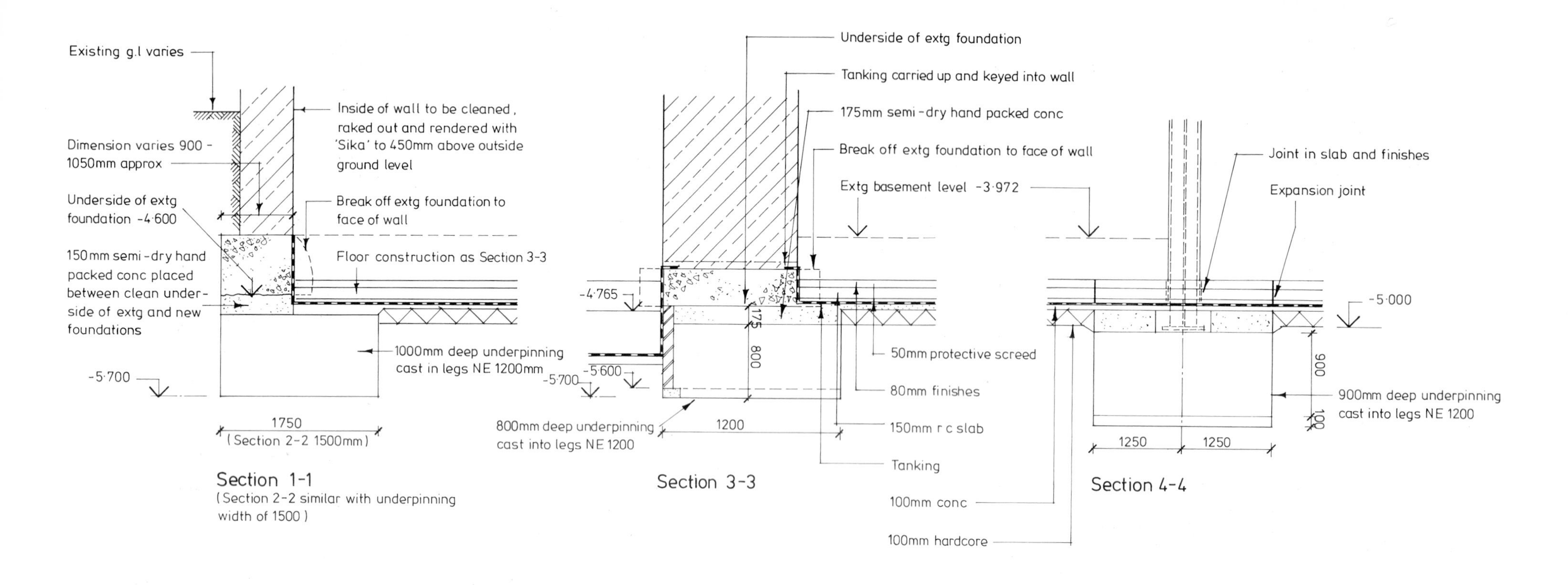

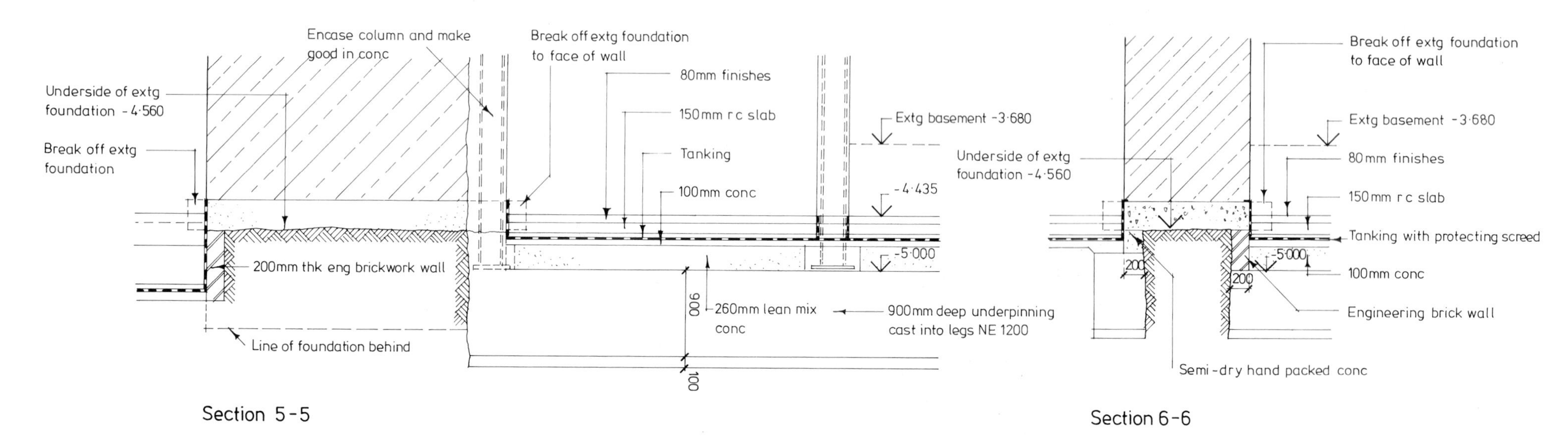

Figure 7.71 3–13, George Street, Edinburgh: foundation sections.

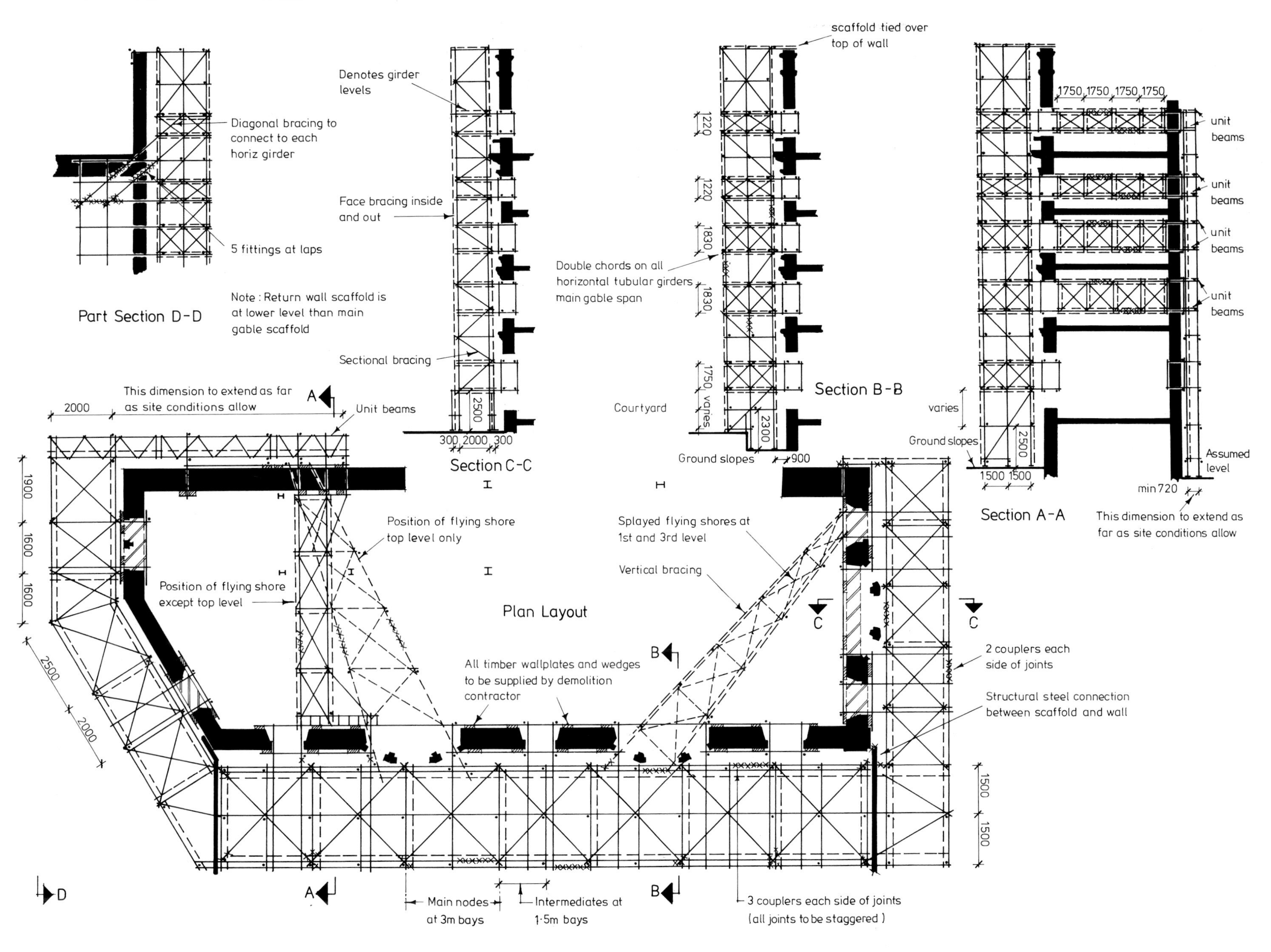

Figure 7.72 3–13, George Street, Edinburgh: temporary support system.

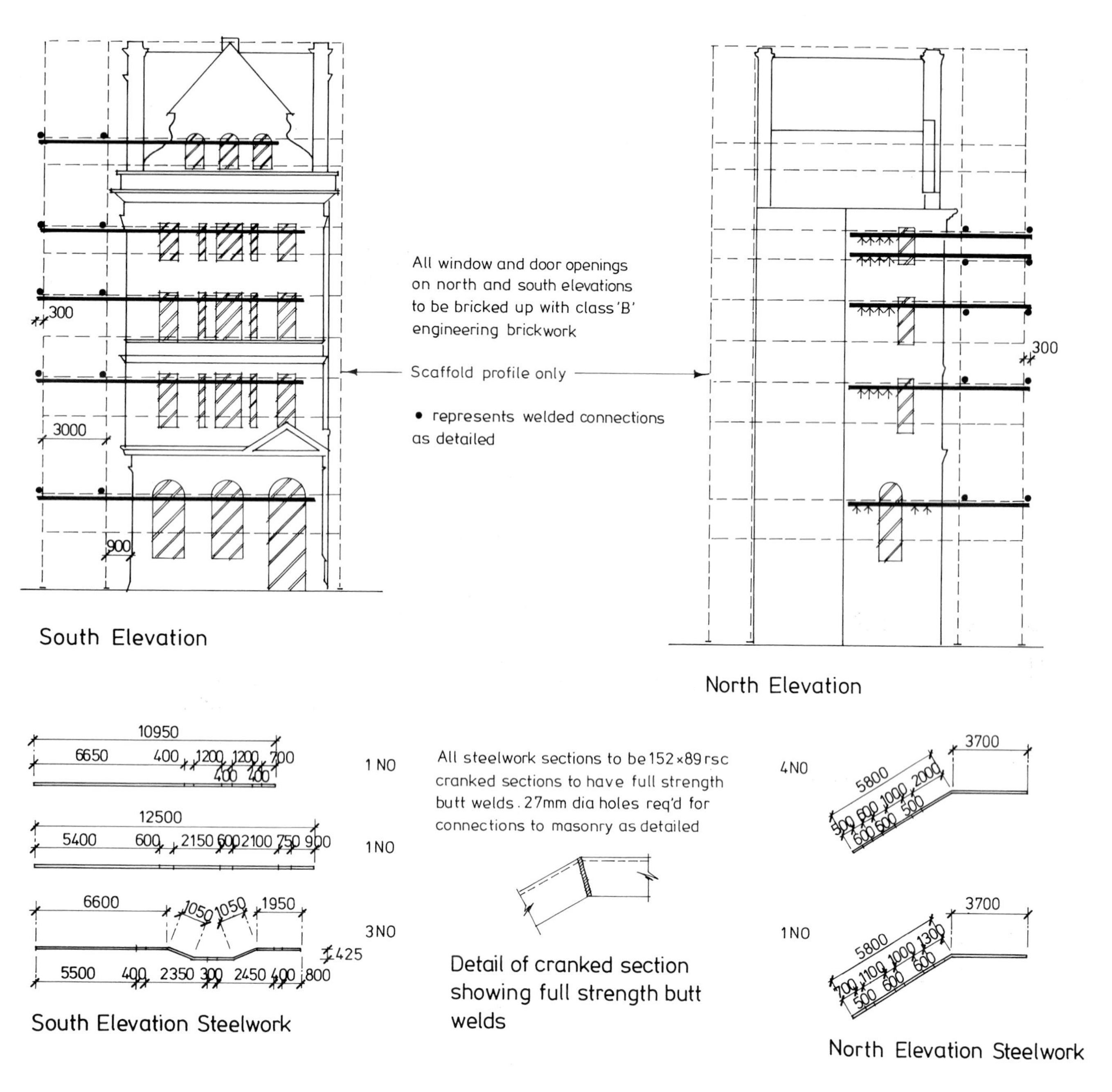

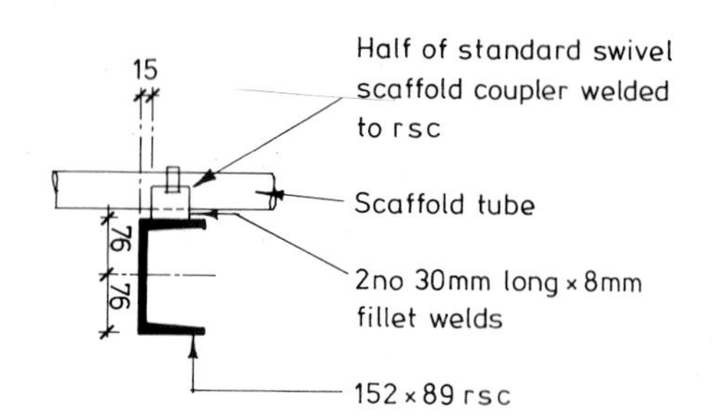

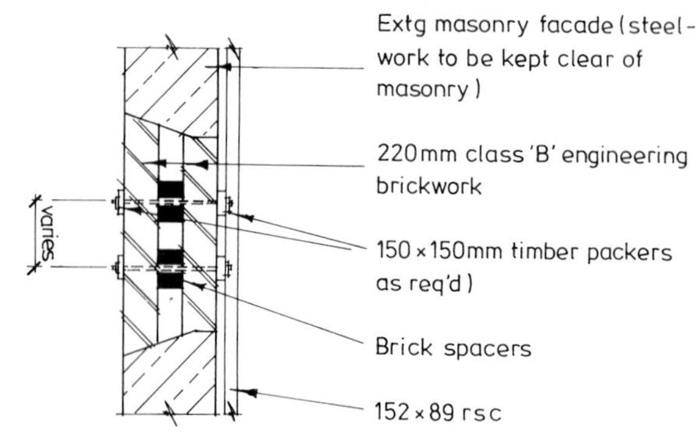

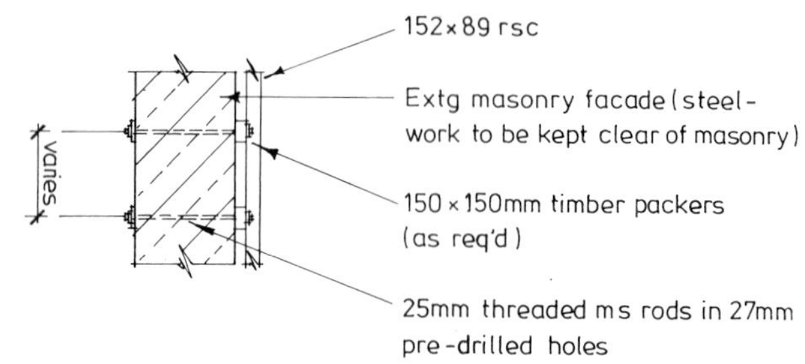

Figure 7.73 3–13, George Street, Edinburgh: temporary support system.

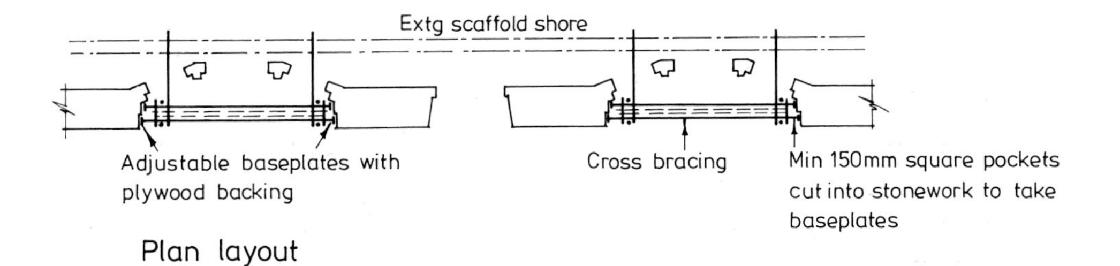

Shoring to bay windows of west elevation

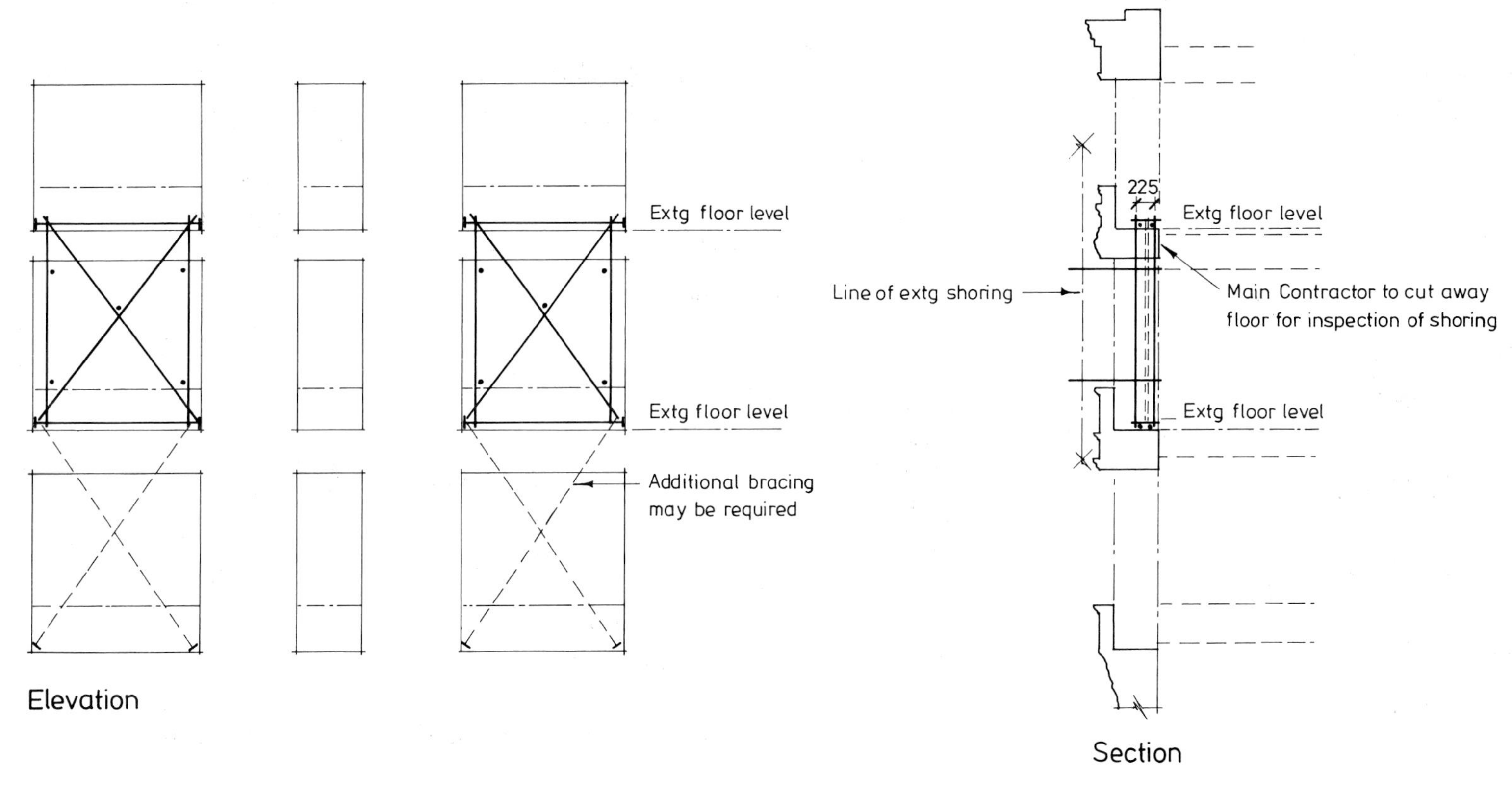

Figure 7.74 3–13, George Street, Edinburgh: temporary support system.

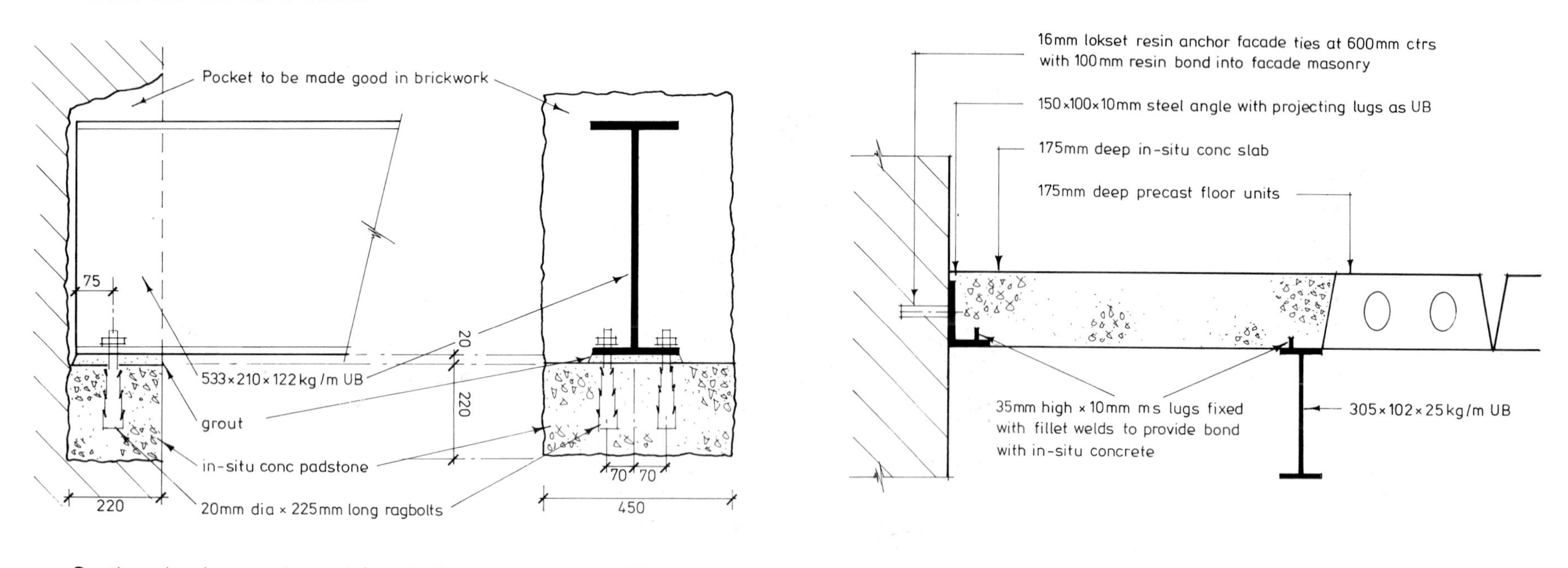

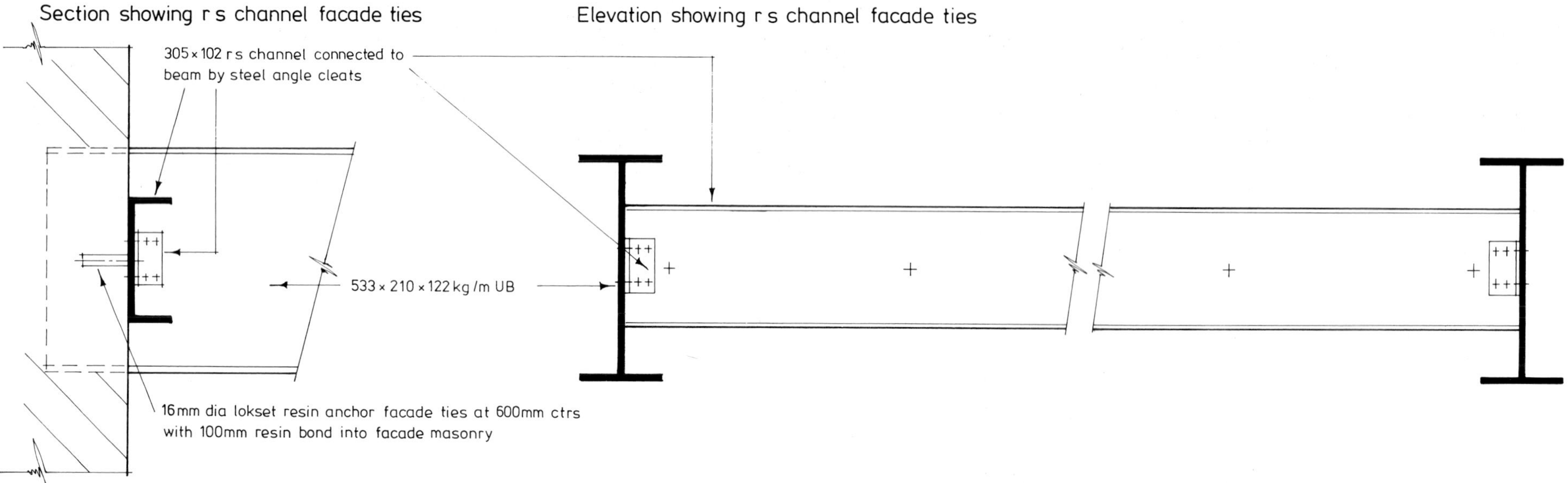

Figure 7.75 3–13, George Street, Edinburgh: facade-tie details.

Figure 7.76 3–13, George Street, Edinburgh: before redevelopment.

Figure 7.77 3–13, George Street, Edinburgh: after redevelopment.

Figure 7.78 3–13, George Street, Edinburgh: retained facade near completion of demolition works showing elements of temporary support system.

Figure 7.79 3–13, George Street, Edinburgh: front elevation prior to erection of new infill building showing temporary support system.

Case study seven

Abbey House, Glasgow

Background to the scheme

The original Bothwell Buildings, designed by Alexander Kirkland, one of the major 19th century Glasgow architects, were built in 1849 in a prominent location at the junction of Bothwell Street and Hope Street. The design of the buildings was reminiscent of the Florentine and Venetian Renaissance style, the most notable feature being the ornate ground floor arcading to the Bothwell and Hope Street frontages. The interiors of the buildings had suffered little alteration during their lifetime; the principal alteration having affected the exterior, where much of the ground floor arcading had been removed to allow the insertion of large shop windows, depriving the building of an important element of its character. The buildings comprised basements and three storeys, surmounted by a balustraded parapet and a plain pitched roof with four chimneys. Figures 7.80 and 7.87 show the buildings before the facade retention scheme commenced.

The re-development involved demolition of the whole of the building's interior, leaving only the Bothwell Street and Hope Street facades, behind which was erected a new structure to provide modern office accommodation. The new structure comprised an *in situ* reinforced concrete frame on piled foundations, with *in situ* reinforced concrete tee-beam floors.

In common with the majority of facade retention schemes, additional floor levels were added, the original four (including the basement) being increased to six, this being made possible by the addition of a new roof structure extending to a higher level than the original.

A further important feature of the scheme was the complete reinstatement of the original arcading at ground floor level where it had been replaced by large shopfronts. Because reinstatement in natural stone was considered to be economically prohibitive, glass reinforced concrete was used to 're-create' the original arches, pilasters and ornate capitals. After fixing to the new structure, the glass reinforced concrete castings were finished with a textured stone coating, which was also applied to the existing arcading, in order to complete the restoration of this important and prominent facade. Figures 7.80, 7.81, 7.82 and 7.88 show elevations, a plan and section of the completed scheme.

Temporary support system

The Abbey House facade was supported by a part external/part internal system comprising two independent structural steel frameworks, one to each of the retained elevations. The overall arrangement is shown in Figure 7.83. The supports to both elevations were based upon steel gantries, constructed from large universal section members, which, in turn, carried steel frameworks, fabricated from angle sections, giving direct support to the walls. Although the principles used to support both elevations were similar, the designs of their frameworks differed and are therefore described separately.

Bothwell Street elevation

The wide pavement adjacent to the Bothwell Street elevation allowed the steel gantry to be erected with both legs outside the existing building with no encroachment into the roadway. This enabled the gantry to 'straddle' the pavement, therefore allowing it to remain open and ruling out the need to divert pedestrians. The steel framework used to provide direct support to the facade was a compound structure constructed mainly from back-to-back angle sections, this being sprung off the gantry to form a cantilevered section which enveloped the facade. This cantilevered section of the framework in turn supported pairs of horizontal 'I' section walings on each side of, and parallel to, the wall at three levels. These pairs of walings acted as collars between which the facade was rigidly secured by means of timber folding wedges. Two important considerations in the design of the supports to the Bothwell Street elevation were the high eccentricity of loading on the gantry caused by the cantilevered framework sprung off it, together with the tensile and compressive stresses in the gantry's legs caused by wind loading on both sides of the exposed facade. The supports to the Bothwell Street elevation are shown in Figure 7.84.

Hope Street elevation

Because of the much narrower pavement adjacent to this elevation, and the greater density of traffic along Hope Street, half the width of the steel gantry had to be located inside the building, in order to minimize encroachment into the roadway. This necessitated locating the gantry's outer legs in the pavement and its inner legs inside the building, with its beams passing through the wall. The obstruction of the pavement inevitably resulted in diversion of pedestrians, partial obstruction of the road and minor interference with traffic flow along Hope Street. The gantry's position, however, enabled the steel framework to be sprung directly off it and to envelope the facade without the need for a cantilevered section. The framework supported the Hope Street facade in an identical manner to that used for the Bothwell Street facade, employing pairs of horizontal walings and timber folding wedges at three levels. The Hope Street elevation's support system is shown in Figure 7.85.

Facade ties

The retained facade was tied back to the new structure by means of steel angles bolted to the upper surfaces of the floor slabs and resin anchored to the facade masonry. One such connection was located between each window opening at every floor level. The exact location of each connection was not specified so that the position where the masonry was most sound could be selected in order to effect the best possible anchorage. Facade tie details are shown in Figure 7.86, and the sequence of operations for their installation was as follows.

1. Two 300 mm long × 20 mm diameter bolts cast into top of *in situ* concrete floor slab, with threaded ends left projecting.
2. 400 mm deep × 50 mm diameter hole drilled into facade masonry to take 'Lokset-Rebar'.
3. 150 × 150 × 10 mm steel angle, 400 mm long, with two 75 mm slotted holes in horizontal leg and one 75 mm central slotted hole in vertical leg, bolted to upper surface of floor slab using two bolts previously cast in. When fixed, vertical leg of angle to be in contact with wall, and central slotted hole to coincide with hole in masonry.
4. CBP 'Lokset' type 'P' pre-mixed resinous mortar pumped into hole in masonry via slotted hole in vertical leg of steel angle. This type of pre-mixed resinous mortar was also used, in preference to resin cartridges, at St Paul's House, Leeds (Case study two, p. 59) and at 144 West George Street, Glasgow (Case study four, p. 86).
5. 25 mm diameter 'Lokset-Rebar' tie bar pushed into mortar-packed hole through 75 mm slotted hole in steel angle and secured with locknuts.

Differential settlement

Vertical differential movement between the new structure and the retained facade was allowed for by designing the facade ties to permit such movement, and also by ensuring that no physical bond was formed between the new floor slab edges and the inside face of the wall. At each facade tie connection, the 25 mm diameter resin-anchored tie bar was located at the bottom of its 75 mm slotted hole in the steel angle, and the securing nuts were only hand tightened. This allowed the steel angles to move downwards together with the floors in the event of settlement taking place, enabling the new structure to settle by up to 50 mm without risk of damage to the facade ties, the existing masonry or the new floor slabs. This principle was also used at St Paul's House, Leeds (Case study two, p. 59) and is described in detail in section 6.4. At the interface of the new and existing structures, where the new floor slab edges met the facade masonry, a strip of 12 mm thick fibreboard was inserted to prevent the formation of a bond, therefore allowing differential movement to take place without causing damage.

Foundation design

The design of the new structure required that nine columns be located close to the retained facade, and thus precautions had to be taken in order to minimize disturbance to the existing foundations. The first precaution, taken before the new foundations were constructed, was to underpin the existing foundations using the Fondedile Pali-Radice system (Figure 7.82). This system involves forming small diameter piles through the actual foundations of the structure to be underpinned, and into the subsoil beneath. After the small diameter boreholes have been formed through the structure and the sub-strata, using a special boring rig with drills capable of penetrating masonry, steel reinforcement is inserted and cement/sand grout pumped under pressure into the boreholes to form the piles. The advantages of the system are that, unlike conventional underpinning, it does not require any undermining of the existing foundations, and it does not subject them or the existing structure to harmful vibrations. Following this strengthening and lowering of the existing foundations, the second precaution was to found the new structure's columns on bored piles, the caps of which were positioned approximately 1 m away from the facade's foundations and only 1.5 m deeper. This ensured that potentially harmful vibrations were kept to a minimum and that undermining of the newly underpinned existing foundations was negligible.

Reinstatement of original arcading to the ground floor frontage using glass reinforced concrete

Glass reinforced concrete is a relatively recent and significant development in the field of concrete technology and its special characteristics include the ability to produce relatively thin precast sections of very high strength. The use of this characteristic to produce thin cladding units has become fairly widespread, but its application to the 're-creation' of ornate stonework in restoration work had never been attempted prior to the Abbey House scheme. The glass reinforced concrete castings, shown after fixing in Figure 7.89, comprised three-piece pilaster sections, pilaster bases, arch sections and ornate capitals. The manufacture of the 30–50 mm thick castings was based upon site measurements of the existing arcading, with the exception of the ornate capitals. In order to re-create the latter's fine detailing, it was necessary to take an *in situ* rubber mould from one of the existing capitals, this being used in the workshop as a basis for producing the glass reinforced concrete castings.

The glass reinforced concrete castings were fixed to new engineering brick piers specially constructed for the purpose and built up to the underside of the existing beams spanning the shopfronts which the reinstated arcading replaced. The castings were fixed by bolting through pre-formed slots designed to conceal the bolt heads on completion. After fixing, the smooth glass reinforced concrete units, together with the existing sections of stone arcading, were finished with a proprietary textured stone coating in order to achieve a uniform texture and appearance to the restored elevations.

Architects The Miller Partnership, Glasgow.

Consulting engineers Blyth & Blyth Associates, Glasgow.

Main contractor Taylor Woodrow Construction (Scotland) Ltd.

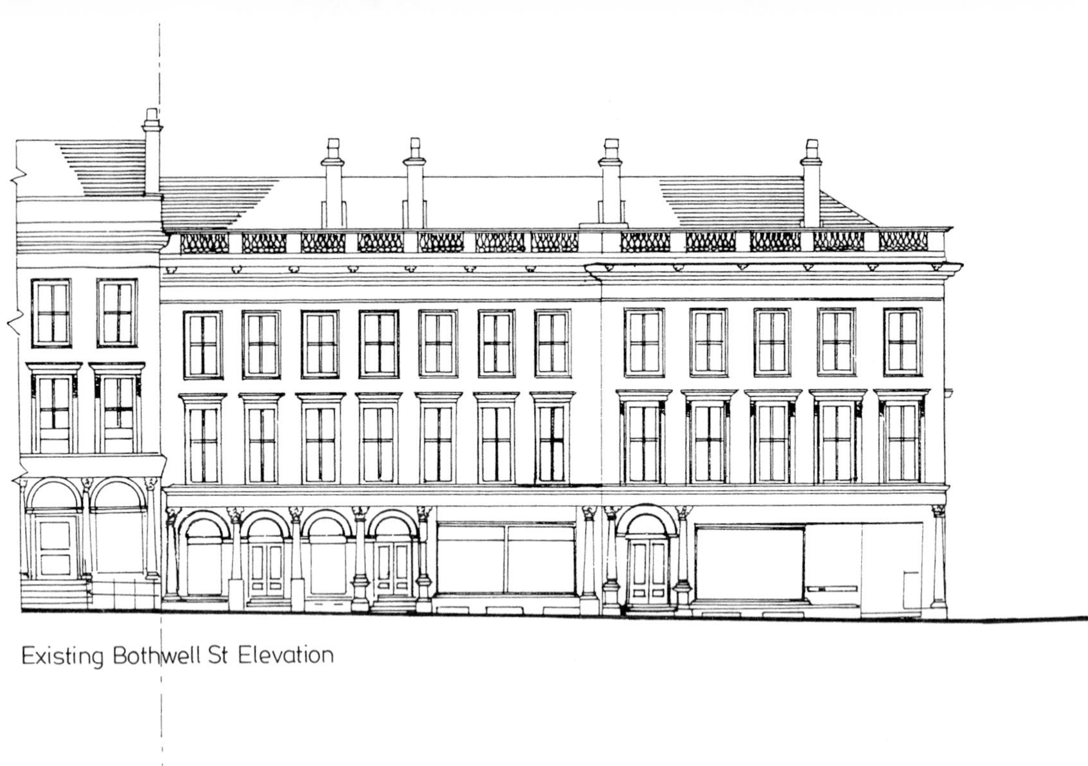

Balustrade to be protected, overhauled straight and true. Where necessary copings and urns to be replaced in precast concrete. All projecting cornices and string courses to be carefully protected and any significant chips in decorative masonry mouldings to be repaired. Cornices to be finished with asphalt weathering and lead drip. All paintwork to existing masonry to be stripped off. 1st and 2nd floor ashlar to be cleaned. All arches and pilasters to Ground Floor which have been removed and replaced with shopfronts to be reinstated with g.r.c replicas to re-create original Ground Floor elevations to Bothwell and Hope St
Existing roof and chimney stacks to be removed and replaced with new roof structure allowing the addition of new 3rd and 4th floors (see Section - fig no 7·82)

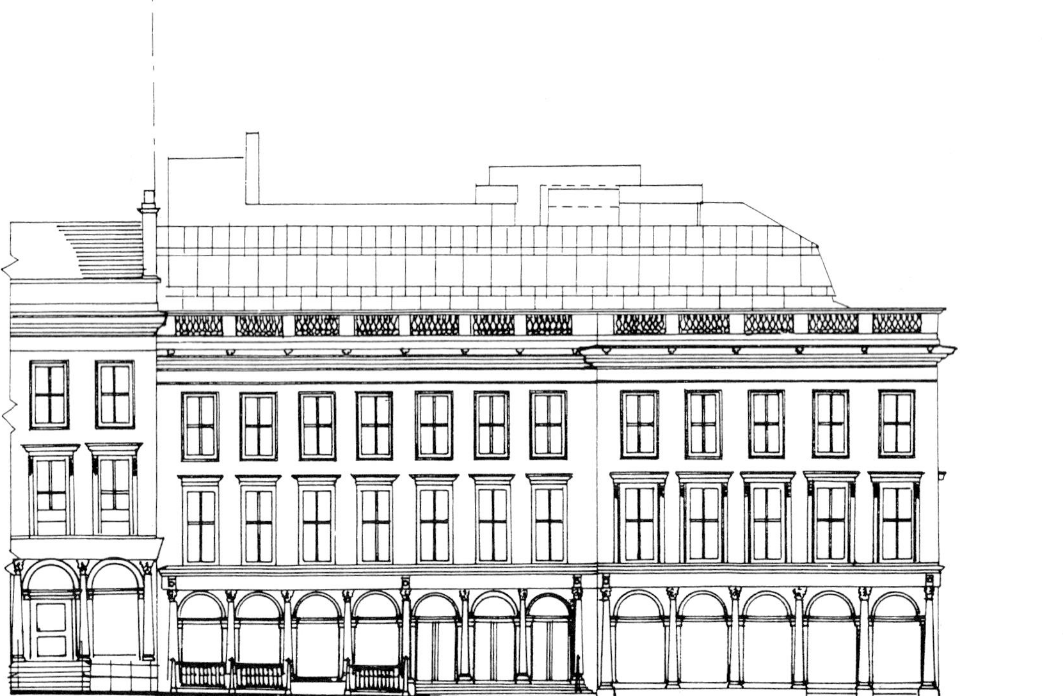

Proposed Bothwell St Elevation

Figure 7.80 Abbey House, Glasgow: Bothwell Street elevation before and after redevelopment.

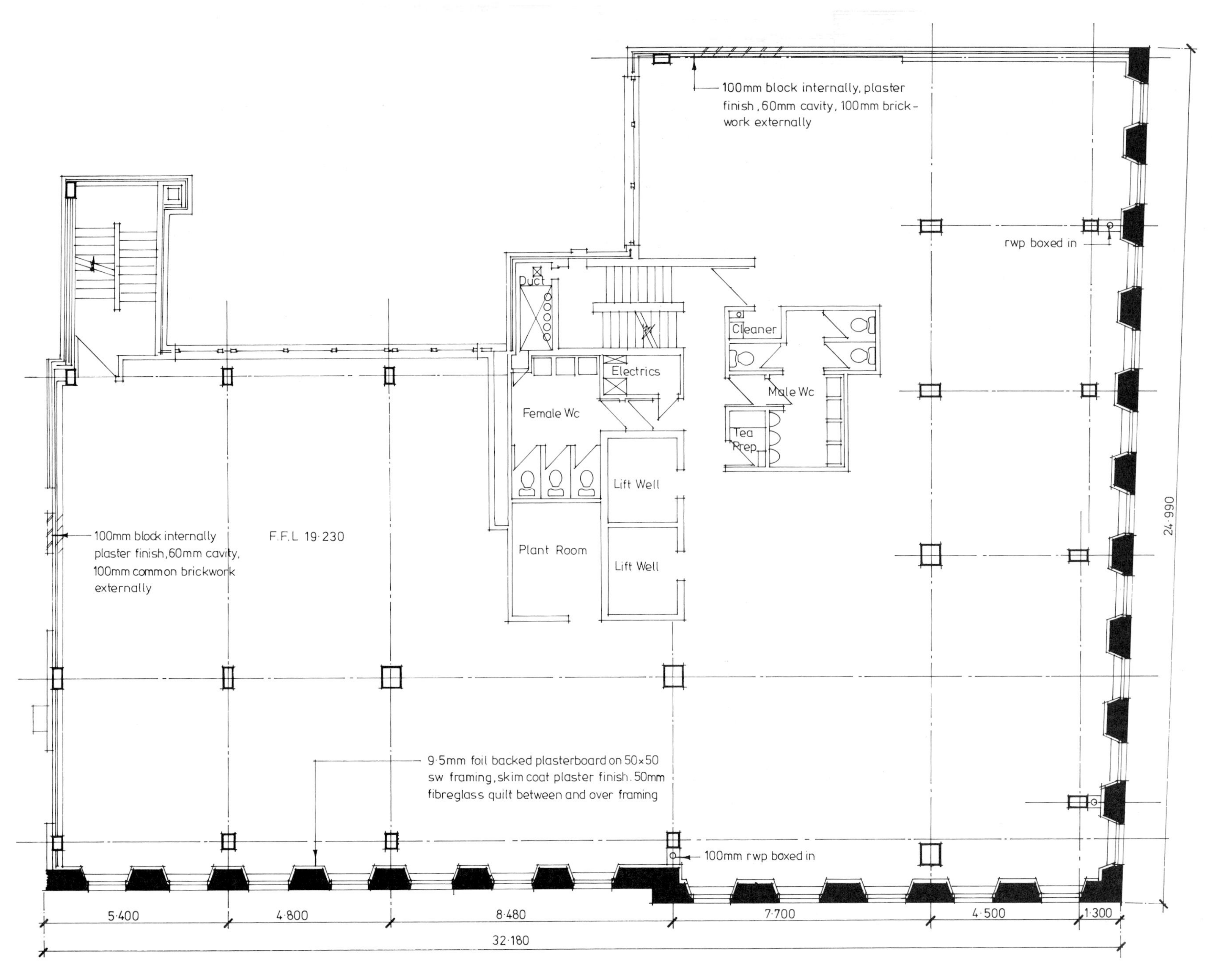

Figure 7.81 Abbey House, Glasgow: first floor plan.

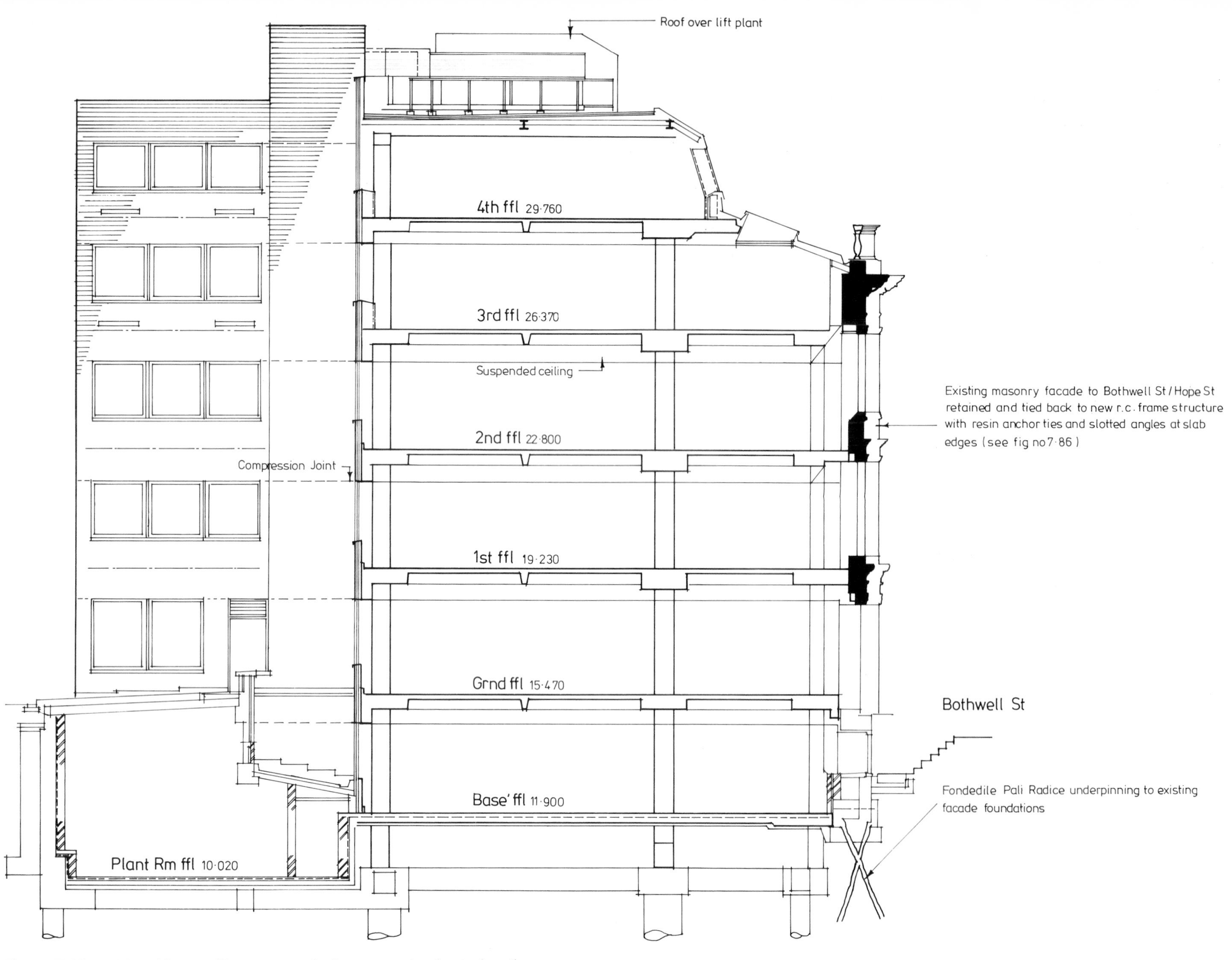

Figure 7.82 Abbey House, Glasgow: vertical cross-section/part elevation.

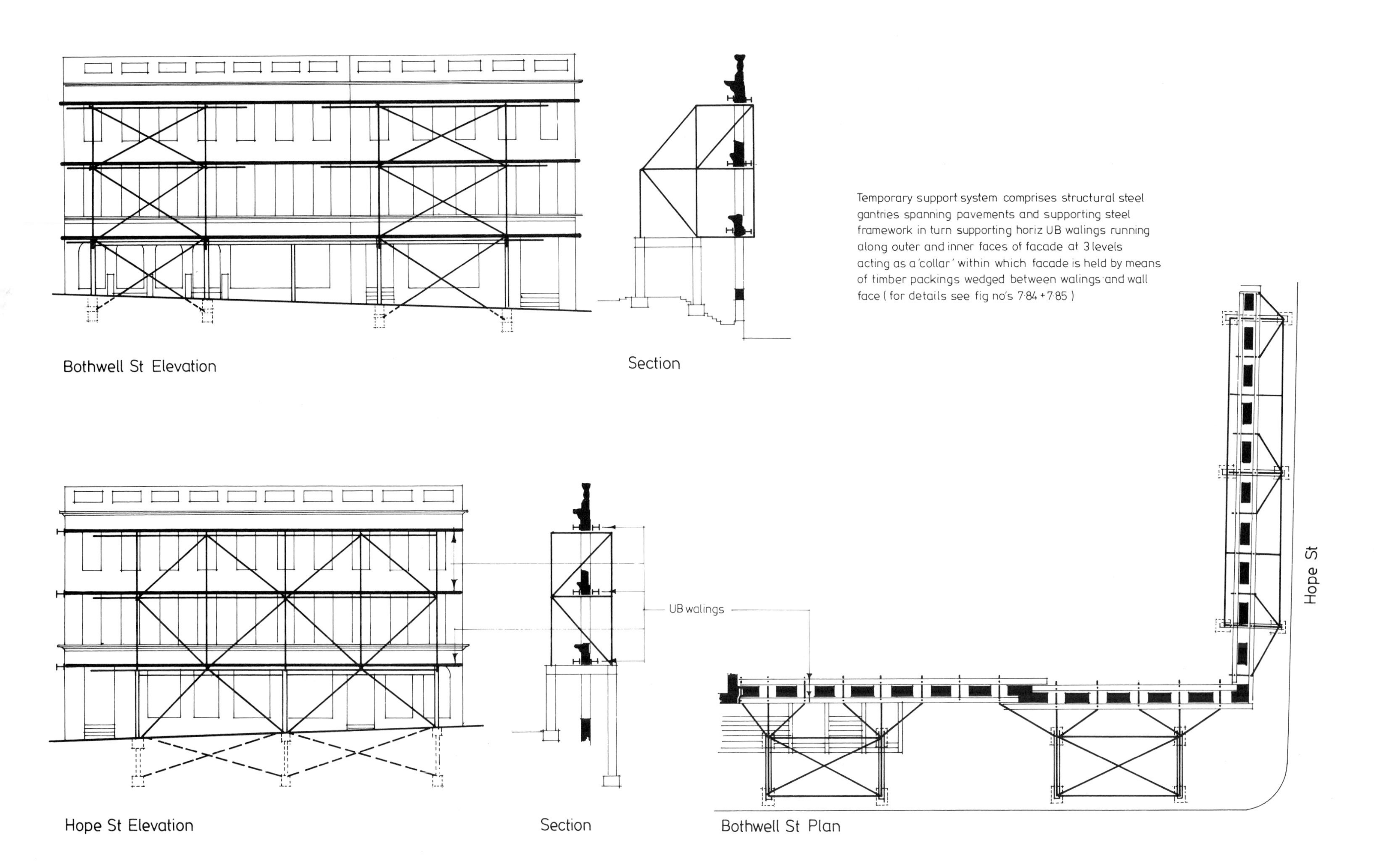

Figure 7.83 Abbey House, Glasgow: temporary support system.

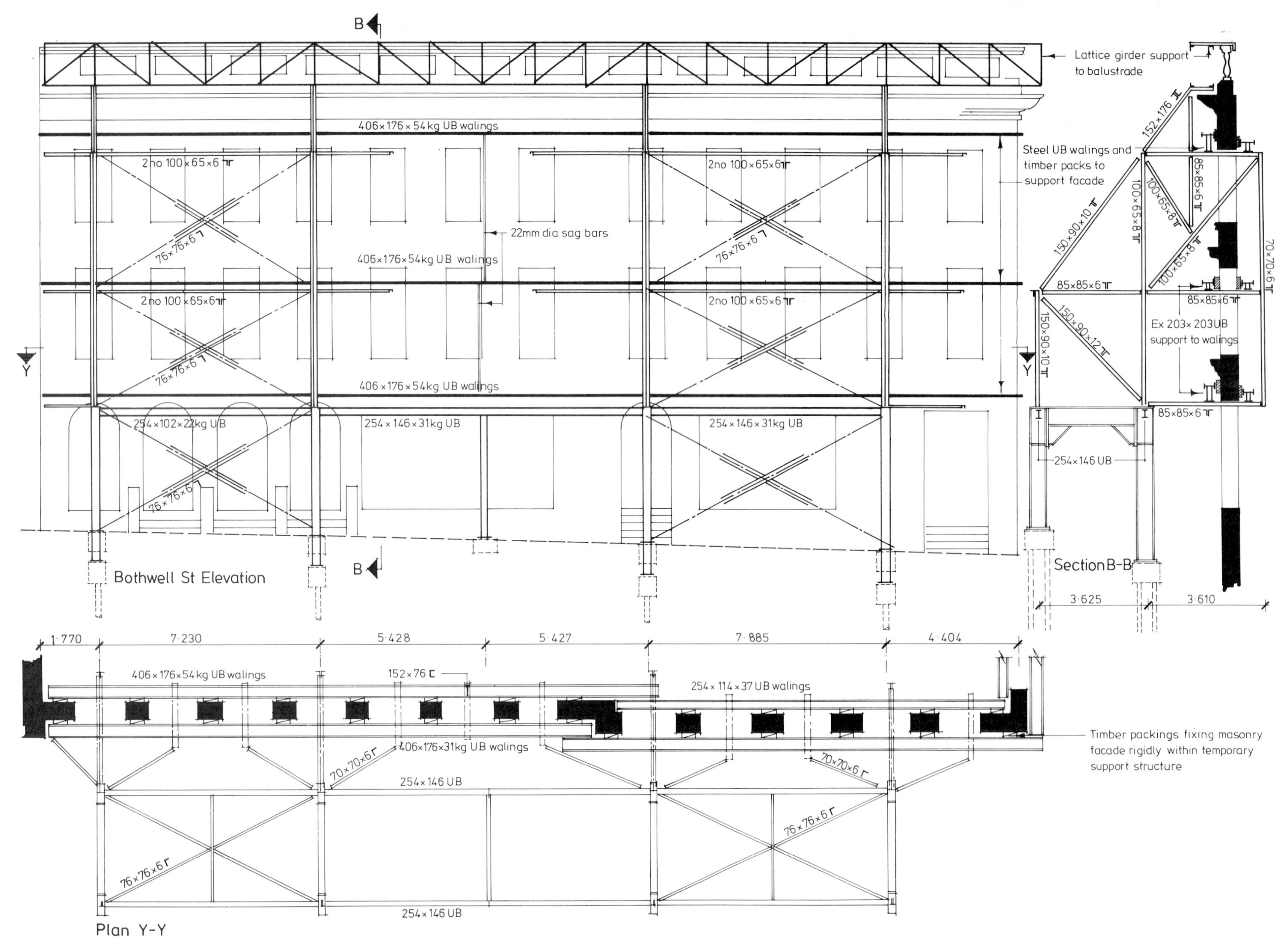

Figure 7.84 Abbey House, Glasgow: temporary support system.

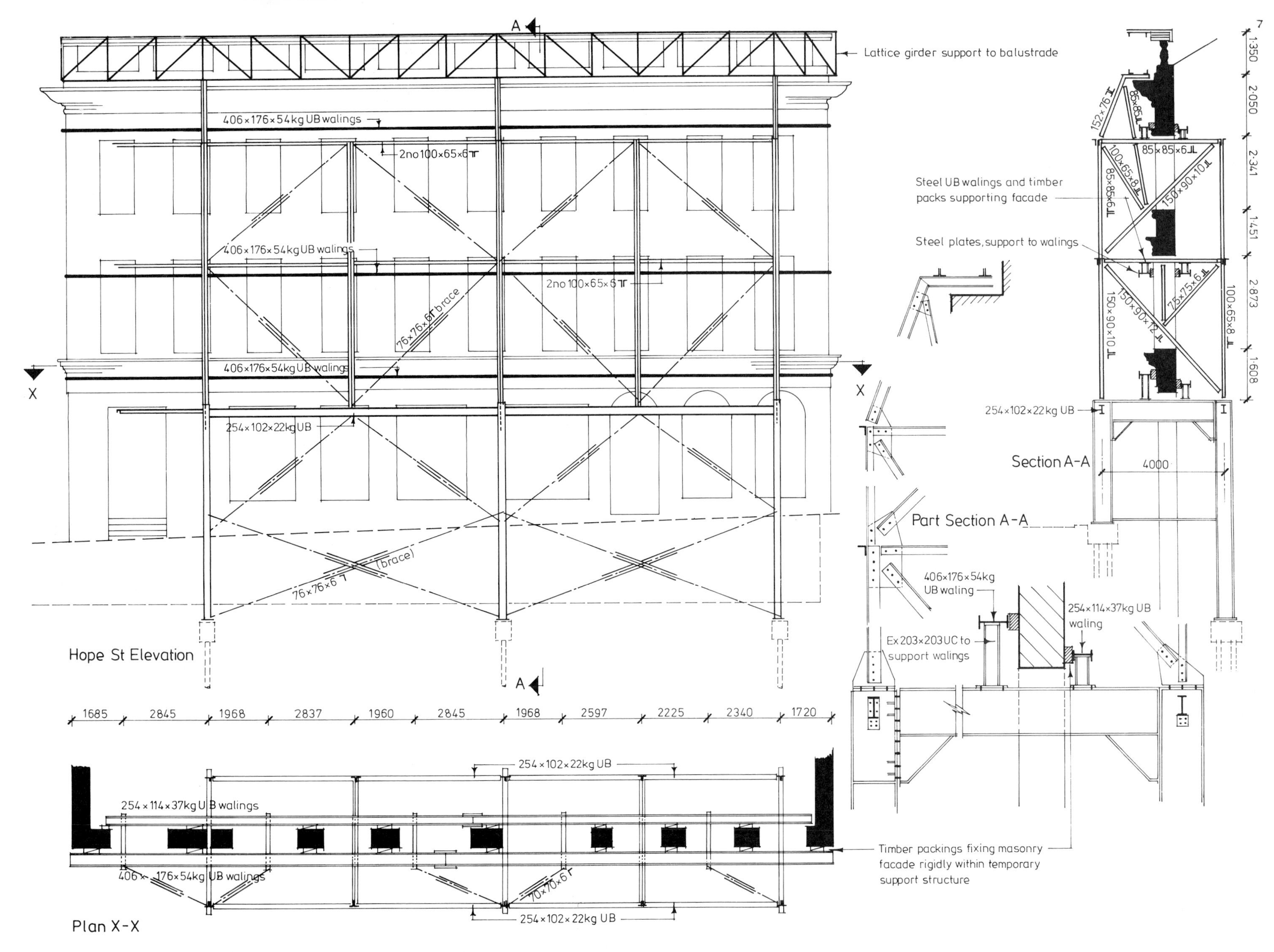

Figure 7.85 Abbey House, Glasgow: temporary support system.

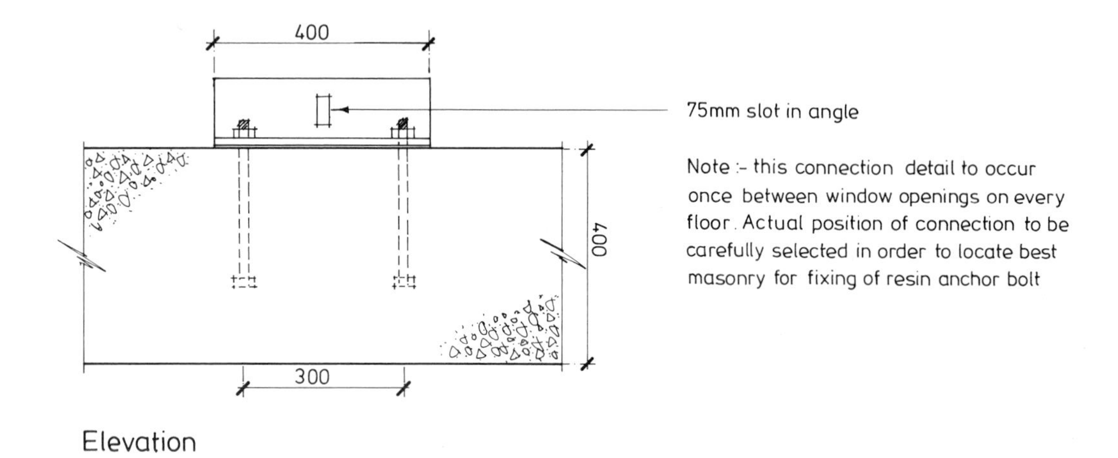

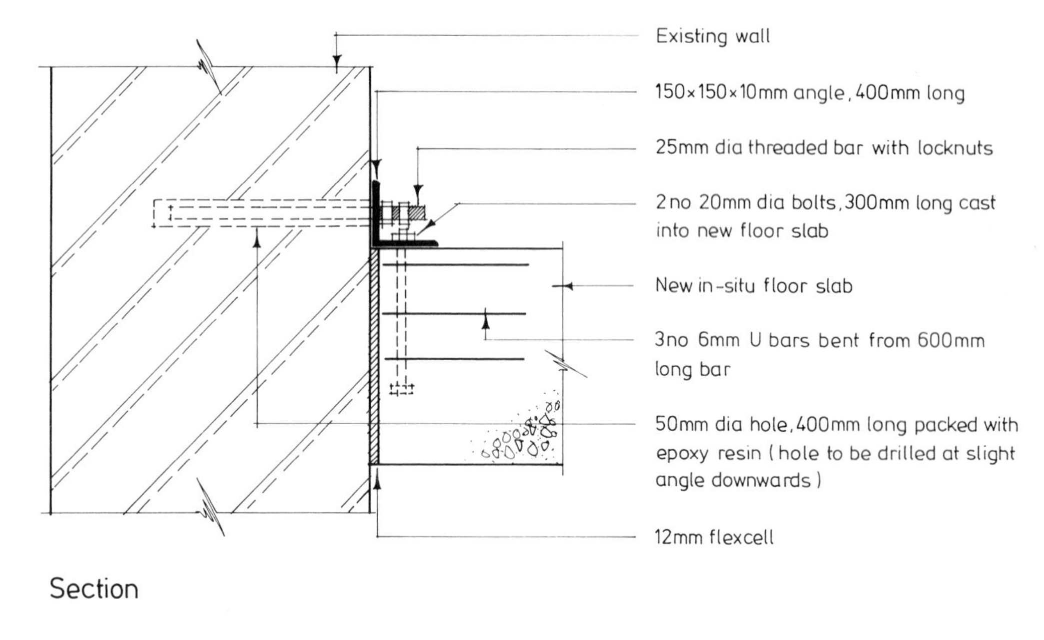

Figure 7.86 Abbey House, Glasgow: facade-tie details.

Figure 7.87 Abbey House, Glasgow: before redevelopment.

Figure 7.88 Abbey House, Glasgow: after redevelopment.

Figure 7.89 Abbey House, Glasgow: new glass reinforced concrete arcading.

Case study eight

Barclays Bank Birmingham

Background to the scheme

The existing four-storey Grade II listed building was constructed in 1867 and formed an integral part of a group of buildings of similar design and period on the north side of Colmore Row, one of the city's main thoroughfares. The stone facade is six windows in width and of Classical design, comprising round-arched windows divided by engaged Doric columns to the first floor; square-headed windows divided by engaged Corinthian columns to the second floor, and paired round-arched windows with pink granite colonettes to the third floor; the facade being surmounted by a modillion cornice. The original ground floor frontage had been altered to form an undistinguished shopfront which was totally out of character with the upper storeys. The shopfront's three plate glass bays were divided by the original cast iron columns which had been retained to continue their structural function and these were the only features of the original ground storey to be retained in the new scheme. The existing front elevation is shown in Figure 7.90.

The building's Classical facade was the principal reason for its listing; the interior, which comprised timber floors supported by timber beams and cast iron columns, possessing no features worthy of merit. The interior structure and layout were also considered unsuitable for adaptation to provide the modern office accommodation that the client required. The re-development therefore involved retention of only the Colmore Row facade, the existing interior and rear elevation, bounded by the party walls with the adjoining buildings, being demolished and replaced by a new structure which provided the same number of floors as the original building. This allowed the modern, air-conditioned office accommodation to be provided without adversely affecting the listed facade and its valuable contribution to its group and the townscape in this part of Colmore Row. Figures 7.91, 7.96 and 7.97 show the front elevation after completion of the facade retention scheme.

The new structure comprises a steel frame supporting composite concrete floors with 63 mm precast concrete planks acting as permanent formwork, and an 87 mm thick *in situ* structural concrete topping. The steel frame's stanchions were encased in *in situ* concrete and its beams treated with vermiculite/cement spray to provide the necessary fire protection. Externally, the plate glass shopfront at ground level was removed and replaced by a modern six-bay frontage designed to reflect the rhythm of the existing fenestration above. Figures 7.92–7.94 show plans and a longitudinal section of the completed scheme.

Temporary support system

The temporary support system was similar in principle to those used at 118–120 Colmore Row, Birmingham (Case study one, p. 44) and 144 West George Street, Glasgow (Case study four); part of the new building's structural frame being used initially to provide temporary support to the facade during the works. This ruled out the need for a truly temporary support system, thereby significantly reducing the overall cost and complexity of the project. An additional benefit, resulting from using part of the new frame, was that the supports were wholly internal, ruling out any need to block the pavement or hinder traffic flow along Colmore Row.

The main part of the new structure comprised a five-bay structural steel frame (Figures 7.93 and 7.94), the first bay of which was erected, prior to any demolition work, to act as the temporary support structure. The only truly temporary elements were the horizontal diagonal bracing members, shown in Figure 7.93, which were removed at a later stage. This structural steel bay had to be erected and braced, and the facade tied back to it, before any major demolition work was carried out. The first and most difficult stage of the work was the construction of the foundations (shown shaded in Figure 7.92) which involved excavations, formwork erection, steel-fixing and pouring of concrete in the restricted conditions of the existing basement. Following the foundation works, the stanchions were threaded through pockets in the existing structure and secured to the foundations. The framework was completed by erecting the beams and temporary bracing members. The facade was secured back to this structural steel bay using resin-anchor ties, shown in Figure 7.95, which would remain as permanent facade ties on completion of the project. When the resin-anchor ties were complete and the facade secured to its 'temporary' structural steel supporting structure, the existing building was demolished and the remainder of the new steel frame erected.

Facade ties

The facade was tied to the new structural steel frame at each floor level using resin-anchor ties which were initially used to secure the facade to its temporary support system, and subsequently to provide permanent lateral ties between the facade and the completed structure.

The ties, shown in Figure 7.95, comprised 762 mm long compound steel angles, bolted to the flanges of the new structure's front beams, and resin-anchored to the facade masonry. The angles, fabricated from steel flats and stiffeners, were bolted to the beams using three 19 mm diameter bolts, and resin-anchored to the facade using two 19 mm diameter tie-bars embedded in the masonry to a depth of at least 225 mm. The procedure used for installing the ties was as follows.

1. Two 25 mm diameter holes drilled into facade masonry at 610 mm centres. Depth of holes to suit thickness of existing masonry, but to be at least 225 mm.
2. 22 mm diameter Celtite 'Selfix' resin capsules inserted into each hole.
3. Two 19 mm diameter Celtite 'Selfix' tie-bars spun into resin capsules inside holes using power tool. This operation breaks the capsule and mixes its ingredients to form a rapid-setting resinous mortar.
4. Resinous mortar left to harden, anchoring the tie-bars into the facade masonry and leaving the threaded outer ends of the bars projecting from the wall face.
5. Vertical leg of compound steel angle, with two 21 mm diameter holes 610 mm apart, located over projecting tie-bars and secured using nuts.
6. Horizontal leg of compound steel angle, with three 21 mm slotted holes at 305 mm centres, fixed to flange of new steel beam using three 19 mm diameter bolts.

Nine resin-anchored ties were installed in this manner at the second, third and fourth floor levels. At first floor level the new slab edge coincided with the existing structural steel beams and column heads supporting the facade, and therefore resin-anchored ties could not be used. The ties used at this level, like those at the higher levels, employed compound steel angles bolted to the new steel beams, but were connected to the facade by bolting directly to the existing steel beams as shown in detail B, Figure 7.95.

On completion of the resin-anchored ties, which secured the facade back to the first bay of the new steel frame, the existing building's interior was demolished and the new construction work commenced.

The final operation, after completion of the new steel frame, was to encase the facade tie angles and the beams to which they were connected with *in situ* concrete. This was carried out concurrently with the pouring of the structural concrete topping to the new floor slab and ensured complete integration of the floors, the beams and the facade ties.

Foundation design

The layout of the new structural frame, and the fact that part of it was used to provide the temporary support system, required that four columns be located immediately adjacent to the retained facade. In common with most facade retention schemes, the stability of the facade was questionable and it was therefore essential that it should not be undermined by the new foundations. The foundations to the four front columns had therefore to be constructed inside the line of the facade, and this resulted in their being eccentrically loaded (Figure 7.92). Eccentric loading of foundations creates overturning moments and tendency to failure and this was overcome by using a balanced-base design, the new foundations being structurally connected to the axially loaded bases of an inner line of columns in order to counterbalance the overturning moments. The balanced-base arrangement is shown shaded in Figure 7.92 and the principles are described and illustrated more fully in Chapter 6.

Differential settlement

The subsoil investigations led to the conclusion that the new structure would undergo only minor settlement, subjecting the facade ties and the interface between the new and existing structures to negligible stress. Any provisions for differential settlement in the facade ties, or at the interface between the new and existing structures, were therefore unnecessary and, as the details in Figure 7.95 show, the connections were fully rigid.

Architects Barclays Bank Property Services Department, Birmingham.

Consulting engineers Roy Bolsover & Partners, Birmingham.

Main contractor T. Elvin & Son, Birmingham.

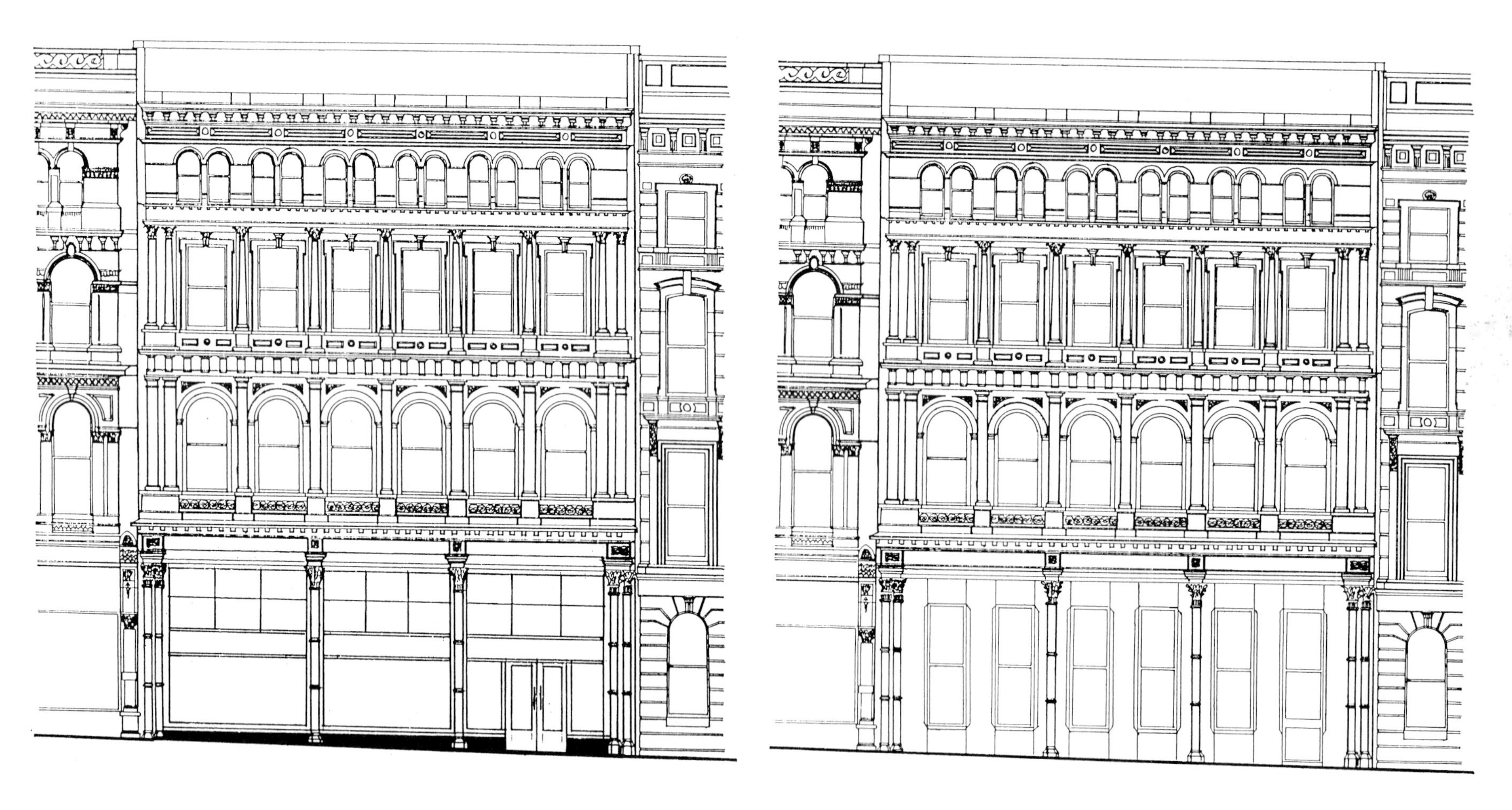

Figure 7.90 Barclays Bank, Birmingham: front elevation before redevelopment.

Figure 7.91 Barclays Bank, Birmingham: front elevation after redevelopment.

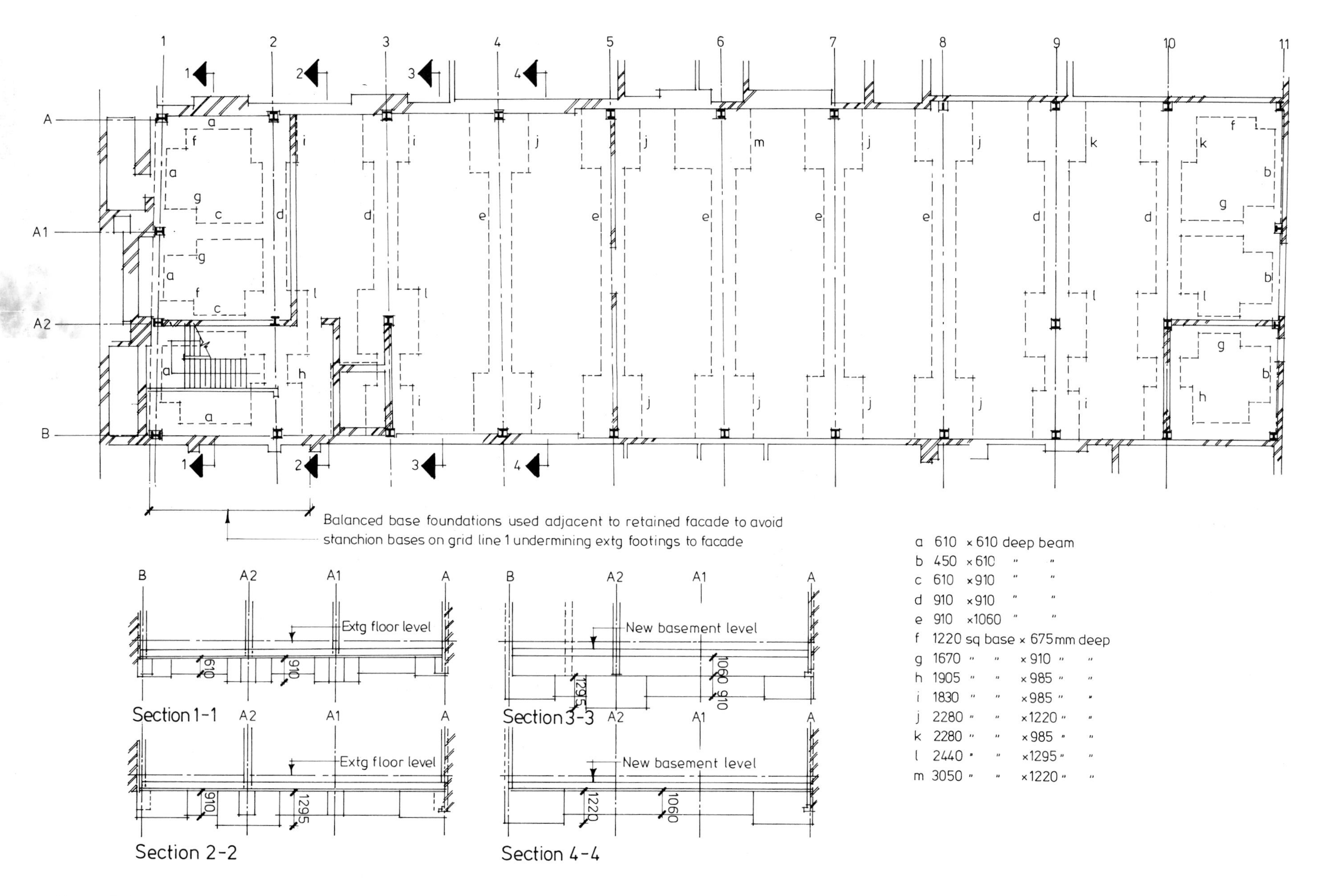

Figure 7.92 Barclays Bank, Birmingham: basement plan and foundations.

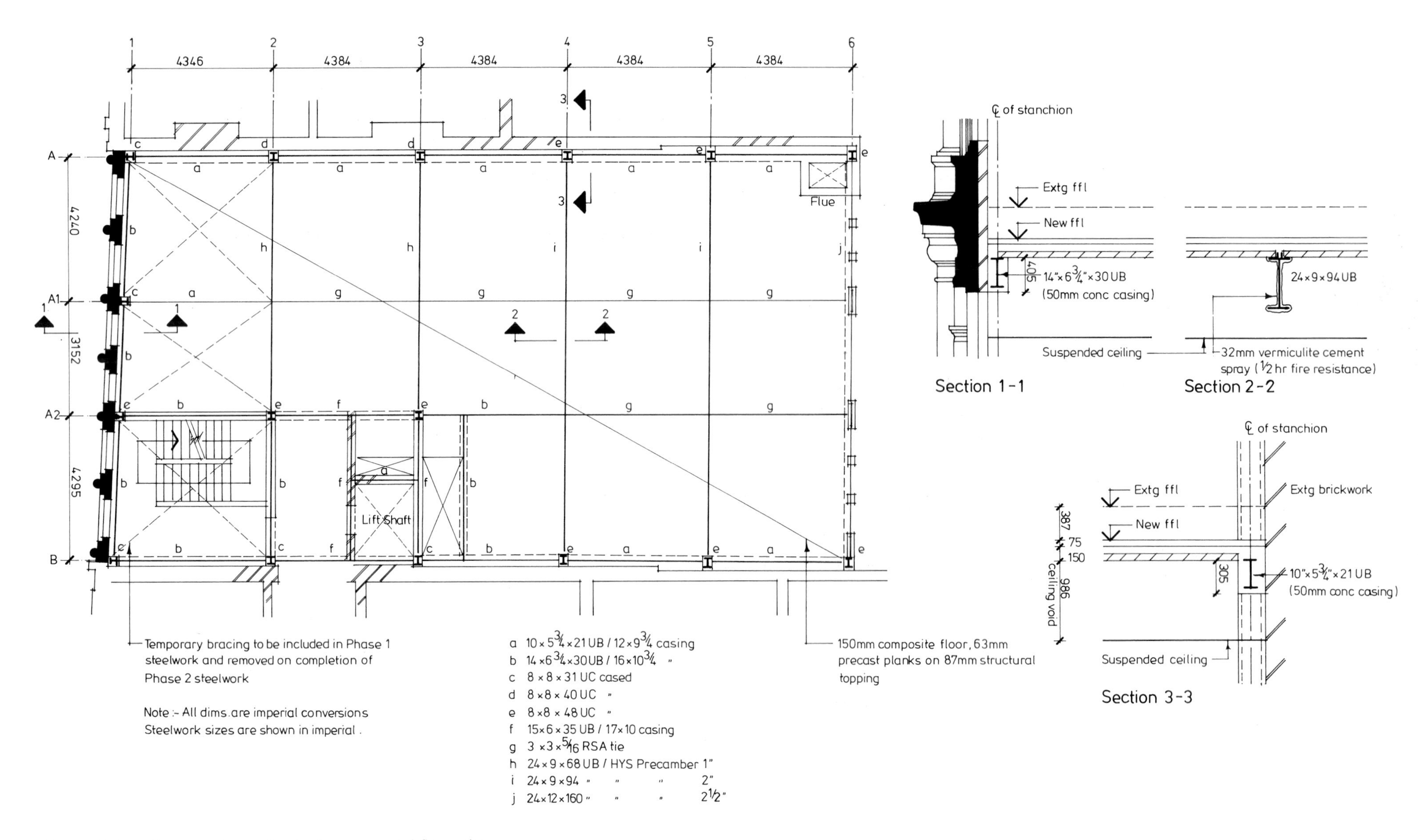

Figure 7.93 Barclays Bank, Birmingham: second floor plan.

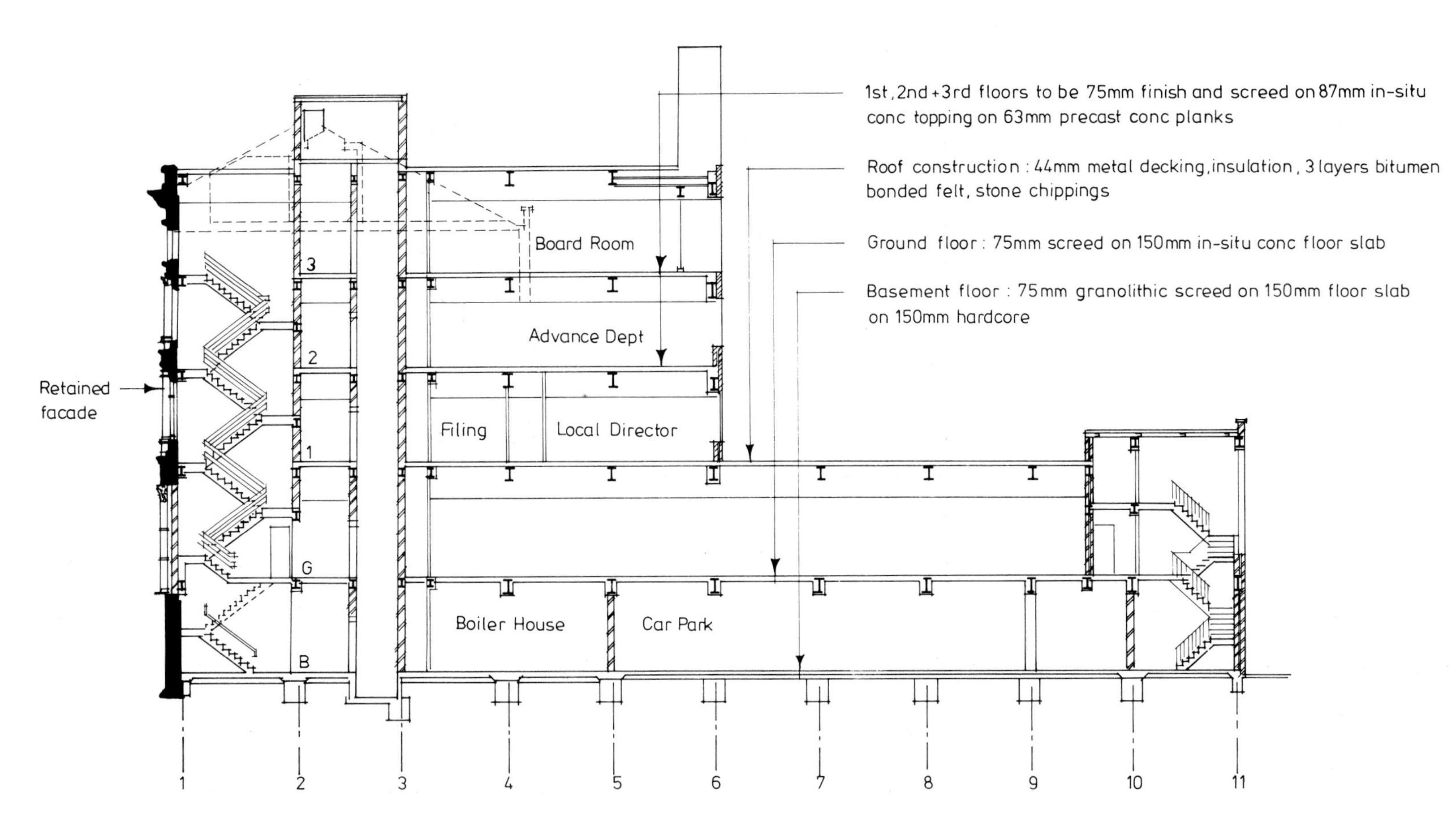

Figure 7.94 Barclays Bank, Birmingham: longitudinal cross-section.

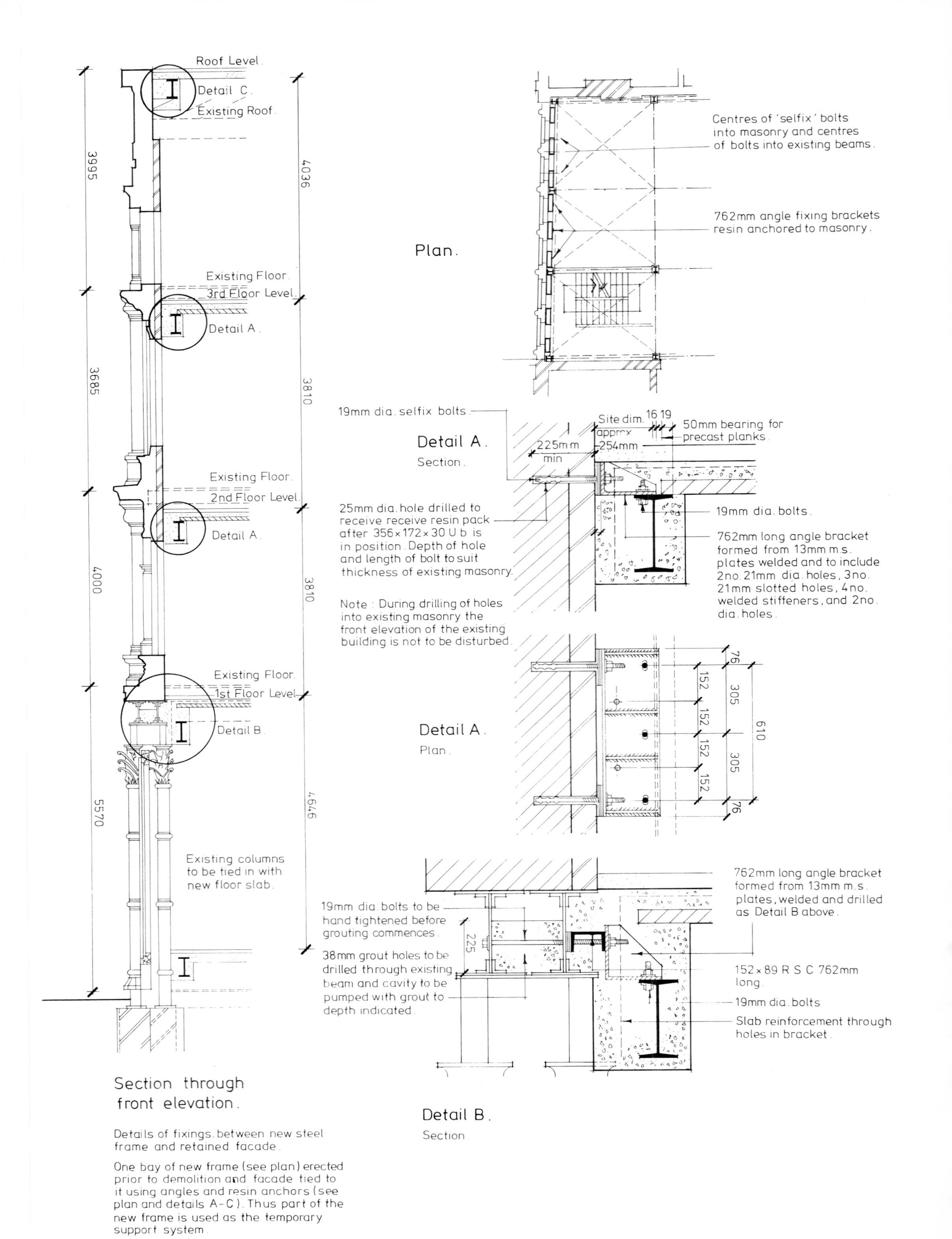

Figure 7.95 Barclays Bank, Birmingham: temporary support and facade-tie details.

Figure 7.96 Barclays Bank, Birmingham: after redevelopment.

Figure 7.97 Barclays Bank, Birmingham: after redevelopment showing the original cast iron columns retained as part of the new ground floor frontage.

Index

UNIVERSITY OF WOLVERHAMPTON
LIBRARY